THE RELEVANCE OF
ALBERT SCHWEITZER

at the Dawn of the 21st Century

Edited by

David C. Miller, M.D.
and
James Pouilliard

University Press of America,® Inc.
Lanham · New York · Oxford

Albert
Schweitzer
Institute

Copyright © 1992, 2000 by
The Albert Schweitzer Institute for the Humanities

University Press of America,® Inc.
4720 Boston Way
Lanham, Maryland 20706

12 Hid's Copse Rd.
Cumnor Hill, Oxford OX2 9JJ

Library of Congress Cataloging-in-Publication Data

The relevance of Albert Schweitzer at the dawn of the 21st
Century / edited by David C. Miller and James Pouilliard.
    p.   cm
Originally published: 1992.
Papers presented at the International Albert Schweitzer
Colloquium held Aug. 23 and 24, 1990 at the United Nations.
Includes bibliographical references.
l. Schweitzer, Albert, 1875-1965—Congresses. I. Title: Relevance
of Albert Schweitzer. II. Miller, David C., 1917-III. Pouilliard, James.
IV. International Albert Schweitzer Colloquium (1990 : United Nations)
    B2430.S374R45   2000   610'.92—dc21   00-062893 CIP

ISBN 0-7618-1834-0 (pbk: alk. ppr.)

♾™ The paper used in this publication meets the minimum
requirements of American National Standard for Information
Sciences—Permanence of Paper for Printed Library Materials,
ANSI Z39.48—1984

To our wives,
Rhena Schweitzer Miller
and
Lauren Chisnall

And to the two individuals
who created the
International Albert Schweitzer Colloquium,
Harold Robles and Bernie Jaroslow

This edition is dedicated
in memory of
David C. Miller, M.D.,
friend of Albert Schweitzer
husband of Rhena Schweitzer Miller
and editor of this publication

Acknowledgments

The preparation of *The Relevance of Albert Schweitzer at the Dawn of the 21st Century* has been the project of a dedicated few who have given generously of their time and talents.

From the outset, we had the benefit of many constructive insights from Rhena Schweitzer Miller and Harold Robles.

Antje Bultmann Lemke oversaw the translation of Dr. Graesser's presentation from the German.

Kelly Fessel read many of the speeches collected in this volume, suggested clarifications, organized the speakers' references and prepared final versions of many of the speeches.

Maribeth Marsico also lent her editorial talents to the preparation of this collection and assumed the huge task of producing the final camera-ready manuscript.

Lauren Chisnall read the manuscript after we thought it was "final" and reminded us that no manuscript can have too many proofreaders.

All photographs in this volume, except those of Bailey, Cousins and Mittermeier, are by Krasner/Trebitz, Cooper Station, New York.

In thanking these friends for their help, we must also accept for ourselves the final responsibility for whatever errors remain.

D.C.M.
J.P.

Contents

Foreword

As a young boy Albert Schweitzer was aware that lives were precious. After many years of scholarship in theology, music, philosophy and medicine, as well as exposure to the countryside of his Alsace homeland and to the jungles of Africa, he distilled his philosophy into three simple but immensely powerful words: "Reverence for Life". It is for this that he wished to be remembered more than for his many accomplishments as organist, writer, hospital builder, physician, humanitarian and voice on the world stage on behalf of peace between peoples.

His legacy was the inspiration for a colloquium in August 1990 held at the United Nations in his honor on the occasion of the twenty-fifth anniversary of his death. Issues with which he was deeply concerned -- peace, healthcare, ethics, human rights, the environment, and above all respect for all life -- were visited and explored. World-renowned peacemakers, philosophers, environmentalists, medical innovators, and Nobel laureates participated as panelists. Their presentations are published here for the benefit of those concerned with ethical behavior and peace in our world.

This volume is republished on the tenth anniversary of the colloquium. The issues discussed are of equal, if not greater concern to the welfare of our planet today. Wars end and others begin; some diseases are controlled while others flourish; natural resources continue to be exploited; children starve and the gap between the haves and have-nots widens. Technological advances accelerate and with them the risk of major disasters as they land in the wrong or careless hands. Albert Schweitzer's message – his writings and the life he lived – remains a vital moral compass for society.

The Albert Schweitzer Institute is deeply grateful to the sponsors and planners of the colloquium who made it possible, to the presenters and participants who gave it life, and to the editors who enabled its legacy. This new edition is dedicated to Dr. David Miller (1917 - 1997), who co-edited the first edition with James Pouilliard and served with his wife Rhena Schweitzer Miller as honorary co-chair of the colloquium. A friend of Albert Schweitzer's, he worked in Lambaréné and devoted his life to issues of healthcare for the under-served.

May the principle of Reverence for Life and the thoughts expressed by the colloquium's presenters serve as a guidepost to individuals, organizations and governments in the 21st century.

Donald Croteau
Chairman
Nikki de L. Lindberg
Board member
Albert Schweitzer Institute
April, 2000

Introduction

In 1935 my father wrote, "There must indeed arise a philosophy more profound and more living than our own, and endowed with greater spiritual and ethical force. In this terrible period through which mankind is passing, from the East and from the West, we must all keep a lookout for the coming of this more perfect and more powerful form of thought which will conquer the hearts of individuals and compel whole peoples to acknowledge its sway. It is for this that we must strive."

Many years have since passed. There were the years of war and mass destruction, followed by efforts of reconstruction. But still inhumanity prevailed in too many parts of the world. Recently the reconciliation between East and West has brought rays of hope, but now again we live in very troubled times, and we still have not found the greater spiritual and ethical philosophy which is so needed.

My hope is that the deliberations of these next two days will bring us some new insights, and may also help us to find better ways to respond to the needs of this world and all who live in it.

At the end of his Nobel Peace Prize address, my father said, "May those who have in their hands the fate of the nations take care to avoid whatever may worsen our situation and make it more dangerous. And may they take to heart the words of the Apostle Paul: 'If it be possible, as much as lieth in you, live peaceably with all men.' His words are valid not only for individuals, but for whole nations as well. May the nations, in their efforts to keep peace . . . , go to the farthest limits of possibility, so that the spirit of man shall be given time to develop and grow strong—and time to act."

Thank you all who have come to this Colloquium for the tribute you are paying to my father, and for giving relevance to his ideas for our time, and for the future, with its enormous challenges.

Rhena Schweitzer Miller

Address of Welcome

By Vasiliy S. Safronchuk
Under-Secretary-General of the United Nations

Welcome to the United Nations. We have been looking forward to this Colloquium, and to the insights and inspiration that will come from your discussions over the next two days. The United Nations, and particularly the Peace Studies Unit of the United Nations, are very pleased to serve as your host for this fourth international colloquium in the Visionaries of Peace series.

I also would like to take this opportunity to express, on behalf of the United Nations, our sorrow and grief on the passing of Pearl Bailey. Pearly Bailey was well known in this organization. For a number of years she was a member of the United States delegation to the regular sessions of the General Assembly.

The United Nations' efforts are clearly in keeping with the belief of Albert Schweitzer that the convictions and actions of individuals are essential to the creation of a more peaceful world. The United Nations is also aware that peace is not just a concern of governments; peace must be the responsibility of everyone.

The Secretary-General of the United Nations, Mr. Javier Perez de Cuellar, regrets that he could not be here. However, he asked me to convey to you, on his behalf, the following message:

> I am pleased to greet the participants of this Colloquium dedicated to the ideals and goals of Albert Schweitzer. The co-sponsorship of this inspiring event by the United Nations' University for Peace testifies to the enduring importance of those ideals and goals to the work of the world organization.
>
> As the United Nations takes up greatly changed and expanded

challenges at the end of the 20th century, the great values and principles of humankind, such as you find in the charter, guide and sustain us.

One of those values is the eternal human yearning for peace. As Albert Schweitzer reminded us in his Nobel Prize address, prophets, great thinkers, artists, and men and women of all kind and degree have, in many times and places, looked forward to the reign of peace. The charter of the United Nations is one more expression of this deeply felt desire for peace. It is our first duty to make that desire a reality.

The theme of your Colloquium recognizes that the security of true peace depends not only on the simple ending of conflict, but also on securing the dignity and well-being of the whole human person. Thus, in the charter's words, all humanity may enjoy better standards of life in larger freedom.

To achieve this, as we now know, disarmament and non-proliferation, human rights, protection of the environment, health and other global issues must be addressed in all their complexity.

Your meeting will be an important contribution to that end, and in so doing will also constitute a fitting memorial to the man you honor. I wish you every success in your deliberations.

Now let me say a few words on behalf of the United Nations, and particularly the Peace Studies Unit.

As we begin our consideration of the relevance of Albert Schweitzer for the 21st century, several political trends and recent events clearly demonstrate the continuing significance of Dr. Schweitzer's beliefs. Today, with the opportunity for hindsight, we can see and understand better than ever before his unique and positive vision of the future. Particularly within the context of global political developments in the second half of the 20th century.

Today we are nearing the end of one important phase of history—the period of decolonization and nation-building. Throughout the 1950's and 1960's, the birth of new sovereign states around the world was guided by the United Nations, based on principles of universal rights and the self-determination of peoples. In 1945, there were only three independent African states—Ethiopia, Liberia and South Africa. At the time of its creation, the United Nations had a total membership of 51 Member States. Today, with the declaration of Namibian independence last March, Africa alone now has 51 independent states and the United Nations membership totals 159. In less than 50 years, the face of the globe has been radically transformed. The story of decolonization is one of the major successes of the United Nations.

Dr. Schweitzer personally witnessed the often difficult transition of colony and dependent territory to independent statehood. His own primary contributions were not, however, to this first political stage of development, but to the second stage. His concerns from his earliest years in Lambarene showed his conviction that political or legal frameworks were not sufficient to ensure the freedoms and the minimum standard of life to which all people are entitled. Dr. Schweitzer's own work, beginning in 1913, focused on social development, on the fulfillment of basic human needs.

It took several more decades until the importance of economic and social development was clearly recognized and accepted by the international community. As many new nations struggled for survival, it became clear that political independence could bring little joy for those held in bondage by hunger, disease and poverty. The second phase of development began with awareness of the need to ensure that political frameworks would be complemented and integrated with viable economic and social systems.

In this context, the United Nations launched the First Development Decade in 1961, the first concerted effort to mobilize international action for development. That task has been a major concern of the global community ever since. We are now in the last year of the Third Development Decade, with many goals still to be achieved. Dr. Schweitzer clearly identified social development very early as one of the major priorities of the 20th century. It will no doubt remain a priority into the 21st century.

Although the second stage of development has not yet been completed, we are actually entering the third one. Dr. Schweitzer did not live to see the true beginnings of this new age. However, his perceptions of the future are revealed in his writing. In accepting the Nobel Peace Prize in 1954, Dr. Schweitzer said, "Whether we secure a lasting peace will depend upon the direction taken by individuals—and, therefore, by the nations those individuals collectively compose. This is even more true today than it was in the past."

Today we are entering a period of globalization and increased participation, by people and nations, in world affairs. Peoples in Eastern Europe and the Soviet Union have developed greater political and economic pluralism as they undergo a real revolutionary transformation. The government of South Africa is taking steps toward recognizing the basic human rights of all people. Greater ease of communication allows greater awareness and understanding of other cultures, philosophies and ways of life. New interest has been awakened in using international cooperation as a means of solving problems and achieving common goals.

At the same time, people worldwide are recognizing that many threats to peace result from their own actions—environmental degradation, drug and narcotic abuse, overpopulation, and ethnic and religious strife. The current dramatic events in the Middle East, and particularly in the Gulf area, are a vivid demonstration of how fragile is the world's peace. The United Nations Security Council is now grappling with this situation, and works practically around the clock trying to defuse the gigantic powder keg in the Gulf area. And, I think we would wish it every success in this endeavor.

There is a new awareness that governments—and the United Nations—cannot solve these problems alone. Our future depends on the commitment, cooperation and hard work of informed and responsible individuals in every region of the world.

The life of Albert Schweitzer provides a unique perspective on the political developments that have shaped this century. He experienced and contributed to the first two phases of development; his vision for the future outlined the third. The example which he provides us, from both his work and from his personal convictions, will continue to inform our own efforts for some time to come.

August 23, 1990

Arias

Keynote Address
The Struggle for Peace

By Oscar Arias Sanchez

We have been called together to remember, on the twenty-fifth anniversary of his death, an upright man. We are celebrating this occasion in memory of Albert Schweitzer, a man who dedicated his life and knowledge to the well-being of others. We are gathered here today to tell the world that Albert Schweitzer has not disappeared, that death could never be strong enough to eliminate one of the most brilliant and exemplary spirits of our time.

In honoring the memory of a man who made the struggle for peace and human dignity his most sacred mission, nothing is more appropriate than some reflections about the threat that nuclear arsenals pose to life on our planet. We all agree that the destruction of these arsenals is imperative. Maintaining them means not only the waste of vast material resources, but an unprecedented example of foolishness: their destructive power is so great that one small fraction of their content would be enough to extinguish any trace of human life.

No effort should be spared in encouraging the powers that possess nuclear arms to totally destroy them, and to prevent other states from being able to manufacture them. We can be sure that if Albert Schweitzer were alive today, he would consider that to be one of his most important tasks.

However, we can also be sure that this great apostle of human life and dignity would have taken the condemnation of the nuclear arms build-up to its logical conclusion. He would have condemned the arms race without reservation. The build-up of nuclear, chemical or conventional arms constitutes a brutal aggression against life and civilization; we condemn it equally in all its forms. The potential horror of the atomic and chemical arsenal is greater than any imaginable hell. However, we should be mindful that conventional weapons have caused, and continue to cause, death and destruction throughout those regions of the planet we have named "the Third World." The use and trade of conventional weapons condemns millions of human beings to oppression, to poverty, and also to death.

Let us speak, then, of disarmament: of disarmament in every sense of the word. Let us speak of peace, of change, and of development—benefits denied to the people of the Third World by the manufacture, sale and use of arms. Let us speak today of the enormous dividends that individuals and businesses in industrialized countries gain from the sale of arms—the sale of death. Let us speak of the social sacrifice that the purchase of those arms represents for the most destitute countries of the world. Let us speak of the suffering that war adds to this sacrifice. Let us speak of the moral responsibility that all of that poverty and all of that pain gives citizens, businessmen and the leaders of those countries that base part of their prosperity on the unacceptable business of death.

Let us fear nuclear arms. Let us think of them when we speak of disarmament. For several decades, we have commemorated the holocausts of Hiroshima and Nagasaki. Every August, we remember the fateful moment when, for the first time, a light brighter than the sun itself brought instant death or mutilation to hundreds of thousands of human beings. We must continually remind ourselves of that painful memory, so that our determination to put an end to the nuclear threat never yields.

Let us fear chemical weapons. Let us think of the possibility, now not so remote, of entire populations dying beneath clouds of poisonous gases, of children and old people, women and men, suffering from the indiscriminate and heartrending effects of chemical agents. Let us make this fear an incentive to unite our voices in an outcry for the dismantling of chemical arsenals, and for the termination of substances whose mere possession makes men and nations the enemies of humanity.

Let us condemn the build-up of nuclear and chemical arms, but not make the mistake of forgetting that the deaths in Angola, in Afghanistan, in Vietnam, in Nicaragua and El Salvador have been caused by conventional weapons manufactured and supplied by the industrialized world. Let us not forget that almost all of the countries of the Third World spend a considerable part of their resources on the purchase of conventional arms, resources that they should be dedicating to the improvement of their inhabitants' living conditions. There are developing nations that dedicate hundreds of times more resources to their military expenditures than what they invest in the areas of education and health. It is neither logical nor ethical for the prosperity of countries producing and exporting arms to benefit from irresponsible governments of the Third World subjecting their people to poverty and oppression, or exposing them to the tragedies of war.

The arms trade, this business of death, is an obvious example of the hypocrisy and duplicity that often abound in the international community. States that manufactured and accumulated chemical weapons in great quantities, while proclaiming themselves the champions of peace and justice, condemned a third party because it employed those kinds of arms in a regional war and an internal conflict. We deplore the brutal use of chemical weapons against the Kurdish rebels, but we also have the right to ask ourselves, "What is the moral base of a condemnation arising from countries that manufacture and accumulate weapons with the obvious purpose of eventually using them? What country that exports weapons has the moral right to deplore or condemn the limited wars or internal conflicts of other countries?"

An enormous burden must weigh on the consciences of those nations and individuals who gather the dividends of war for their own material gain. I come from a country that has had a positive experience because of what we have called "the dividends of peace." In 1948, under the inspiration and guidance of President Jose Figueres, the people of Costa Rica decided to completely demilitarize. Figueres opted for suppressing the army, and his fellow citizens believed in the courage and viability of that option.

In 1949, our legislators made that act a constitutional principle and now we and our children enjoy the fruits of that decision. Today, international development agencies recognize that Costa Rica has a standard of living comparable to that of industrialized countries. It is universally accepted that the extraordinary advances of my country in the fields of education, health, housing and social welfare are basically due to the fact that we do not dedicate our resources to the purchase of arms. The absence of an army has strengthened the Costa Rican democratic system to become the most consolidated one in Latin America.

These are the dividends of peace. These are the dividends that would be within the grasp of all Third World countries if we did not dedicate a very important part of our resources to the purchase of arms.

As East-West tensions disappear, we hope that the industrialized nations of Asia, North America and Europe will devote more attention and resources to the solution of the great inequalities that exist within the reality of present North-South relations. We hope that the productive capacity of the war industry will not be maintained at the cost of the blood and well-being of our peoples.

We have heard and read with astonishment and concern the complaints of certain sectors of industrialized nations for whom the advent of peace seems to constitute a misfortune and not a blessing. A few claim that, along with peace, disarmament will bring unemployment and poverty to many citizens of the industrialized world. The industrial-military complex is mobilizing to pressure governments against disarmament, claiming that they cannot condemn thousands of war industry workers to unemployment. We fear that powerful businesses manufacturing arms may attempt to delay disarmament and perhaps even encourage new wars.

Undoubtedly, the national economies that fell into the moral anomaly of depending on the sale of arms will have to subject themselves to difficult processes of readjustment that will demand short-term sacrifices. Entire societies will have to be submitted to limitations that will never be as serious or as painful as the ones that the societies of the Third World have suffered for decades. Poor nations have had to deter many of their countrymen from the labors of peace. That has been our sacrifice. Rich nations must be willing to make the sacrifice of deterring many of their citizens from the labors of war. They must take on the challenge of turning swords into plowshares. The tremendous ingenuity and productive capability, which have until now been dedicated to destruction, should be channeled to solving the immense problems threatening our survival as a species: environment, poverty and disease. This is the concept of security which must prevail as we move to the next century.

This will require the consolidation of a new world that is more just, more secure, more humane. Unfortunately, the present events in the Middle East pose a serious threat to world peace. The leader of a totalitarian state [Saddam Hussein of Iraq] has thrown his people into a repudiable war of aggression, and has caused an intensification of the arms race in that region. As a result of this escalation, the only winners will again be the dealers of death. What may emerge is the intensification of an arms race spreading from the Middle East to the rest of the world like fire in a haystack. Such a tragedy could become the tomb of all our hopes for peace, justice, democracy and development.

Lessons can be learned from our past. An experience that has been onerous for the United States in Latin America may be repeated in the Middle East. The United States, the superpower that has always proclaimed freedom and democracy as indispensable requirements for the development of nations and for international coexistence, has too frequently supported dictatorial governments in which it has seen faithful allies willing to defend its economic and strategic interests. Thus they have lost credibility before the countries of Latin America, whose will has been alienated by an inexplicable contradiction. That contradiction allowed the United States to befriend Pinochet, a despot to the right, while financing a war against the Nicaraguan "Sandinistas," arguing that the country was governed by a dictatorship to the left.

Just as yesterday the involvement of the United States with non-democratic governments in Latin America disillusioned large segments of our populations, we fear that its alliance with autocratic governments in the Middle East today will scar its credibility as the primary champion of the struggle for democracy and freedom.

But in spite of the storm clouds that darken the present, I believe that we live in a time of hope. It may be mankind's last opportunity to renounce its hatred and create a future whose struggles will exclude war. Many threats hang over human life; we cannot lose the opportunity to defeat the most absurd of them all: self-destruction.

From personal experience, I know that dialogue and negotiation are the best paths toward the resolution of conflicts. In Central America, this belief made possible a process of pacification that involved leaders of differing ideological leanings. It was dialogue, and not the power of weapons, that permitted us to put an end to the war in Nicaragua and establish the basis for a future full of hope for the young people of that country. Let us hope that in other conflicted regions of the world dialogue will be given a chance to open new paths for peace.

In my struggle for peace in Central America, I learned many things. I learned the value of humility. We usually tend to believe that our own perspective on a particular conflict is the only one, and that the response we devise is always the correct one. But this is not necessarily true. We must be sufficiently humble to accept the possibility that *we may be wrong*, that other perspectives exist, and that alternative solutions may be more appropriate. We must listen carefully to the voices of the people most directly concerned by a conflict, and work with them to devise a solution, rather than trying to impose solutions from the outside.

I also learned the value of prudence. We must never take action just to please the gallery! But most of all, I learned the enormous value of patience in our struggle for peace. The brutal aggression of Iraq against Kuwait has brought Saddam Hussein the scorn and condemnation of the entire world. Time is against him and he is very alone. This is all the more reason why we must be patient, why we must be humble, and why we must be prudent.

Now is the time to scale down military posturing and belligerent rhetoric. Now is the time for us all to accept our moral responsibility to make peace the first priority. In recent months, so many walls of misunderstanding have come crumbling down. Let us capitalize on this new momentum of reasonable international dialogue. Let us together create a world that does not destroy, a world that nurtures the best in all mankind.

As Albert Schweitzer would have done, let us dedicate our efforts and wills to constructing peace and discouraging violence.

Arms Reduction
and
the Nuclear Threat

Marshall

Cousins

Jack

Leaning

Pauling

Moderator's Introduction

By George N. Marshall

Recent history—the ending of the Cold War, the crumbling of the Berlin Wall, and the collapse of the Warsaw Pact—has given us cause for tremendous optimism. But within the past month, this optimism has been abated by the massive mobilization of forces in the Arabian desert following the invasion of Kuwait by Iraq.

This event, and the anniversary earlier this month of the first nuclear bomb dropped on Hiroshima, August 5, 1945, necessarily broaden this panel's concern with the issues of arms reduction and the nuclear threat.

In this United Nations Colloquium on *The Relevance of Albert Schweitzer at the Dawn of the 21st Century* we must recall that Schweitzer's ethical principle of Reverence for Life was conceived during the First World War when he was seeking the root cause of belligerency and conflict, and a new ethical imperative to diminish them. His first concern was the development of an ethic to end war. It is our duty to carry on Schweitzer's concern.

To Make Future Wars Impossible

By Jennifer Leaning

At the close of the Second World War, as the troops came home from ravaged lands, and as the Allied occupation forces settled into Germany and Japan, the United States began to dream about the future.

Within two years, the U.S. formulated the Marshall Plan, announced in June 1947 as a mission "against hunger, poverty, desperation and chaos"[1] and pledged to support the economic reconstruction of war-torn Europe. In the same year, President Truman issued the pledge that came to be known as the Truman Doctrine, declaring the U.S. intent to intervene in Europe and the Mediterranean to protect the seeds of political freedom in nations newly liberated from the Axis powers. And at the end of 1946, Truman announced the formation of the U.S. Atomic Energy Commission, established to manage and oversee the military and civilian uses of the power discovered in the atom bomb.

In the "blinding dawn"[2] of this early postwar period, the U.S. leaders and public crested a wave of euphoria, vision, and unfamiliar world prowess. Finally, after two hideous world wars, it seemed as if the Allies had truly managed to make "the world safe for democracy."[3]

Yet these dreams were already clouded by premonitions. The scientists who made the atom bomb, very quickly but too late, tried to tell the political leaders the awful truth. A short two days after the surrender of Japan, within 11 days of the atomic bombings of Hiroshima and Nagasaki, Robert Oppenheimer, on behalf of the scientific panel advising on the bomb, delivered the following message to President Truman:

> The development, in the years to come, of more effective atomic

weapons, would appear to be a most natural element in any national policy of maintaining our military forces at great strength; nevertheless we have grave doubts that this further development can contribute essentially or permanently to the prevention of war. We believe that the safety of this nation—as opposed to its ability to inflict damage on an enemy power—cannot lie wholly or even primarily in its scientific or technical prowess. *It can be based only on making future wars impossible. It is our unanimous and urgent recommendation to you that, despite the present incomplete exploitation of technical possibilities in this field, all steps be taken, all necessary international arrangements be made, to this end.*[4]

Yet we all know the outlines of post-war history since 1945. Beginning with the failed Baruch Plan in 1946, to the ongoing war in Afghanistan, we have seen events marking hostile points that have brought both the U.S. and the U.S.S.R. to a massive expansion of technical expertise, an enormous escalation in military spending on nuclear weapons, and an evermore arcane arborization of nuclear war-fighting strategies.

The Dangers

During the 35 years, from 1945 to 1980, public understanding about these weapons came very slowly. Research on the consequences of the bombings of Hiroshima and Nagasaki was kept classified under the orders of Occupation and only released in the medical and government literature in the mid-1950's.[5] It was only in these years that the U.S. managers of atomic weapons began to recognize the enormity of possible dangers posed by radioactive fallout, as reports from the Pacific tests of the hydrogen bomb included instances of illness and death among exposed Japanese fisherman and the population of the Marshall Islands.

Technical analysis of the results of surface tests of these thermonuclear weapons required total revision of the main scientific text used by the U.S. Department of Defense, reissued in 1957, and then again in the 1960's. The major psychological study of victims of nuclear weapons, Robert Jay Lifton's *Death in Life: Survivors of Hiroshima*, was not published until 1967.

Against this backdrop of relative silence and true ignorance, the detailed, urgent clarity of Dr. Schweitzer's messages over Radio Oslo in April 1958 ring like a tocsin for the world.[6] As wide as his listening audience was, would that it had been wider.

The Weapons

Because in the breach, absent widespread public opposition, each year of the arms race has brought a tremendous surge in nuclear weapons development and deployment, on the part of both the U.S. and the U.S.S.R.

The arsenals are diverse, evenly matched according to the experts on both sides, and staggeringly redundant. The U.S. has the capacity to explode 16,000 nuclear weapons over the U.S.S.R.; the U.S.S.R. can explode over 11,000 nuclear weapons on the United States. In addition to these strategic weapons, both the U.S. and the

U.S.S.R. have thousands of "theater" or tactical nuclear weapons, integrated within conventional forces, for use against ship convoys, tank formations, and other targets.[7]

Barriers to Understanding

Up until 1980, the American and world public, and much of the world scientific community outside government, had not paid great attention to the threats posed by nuclear weapons or the risks of nuclear war. A major exception to this statement was the public outcry against surface nuclear explosions and resulting global fallout of radiation. This worldwide clamor, joined by Dr. Schweitzer, gained strength in the late 1950's and 1960's, leading to the Limited Nuclear Test-Ban Treaty in 1963, which banned above-ground tests of nuclear weapons. This treaty ironically served to allay public concerns, and the subject of nuclear weapons returned to its peculiar status of high priority within government; low priority in public debate.

Among the many reasons the public allowed itself to remain isolated from the subject of nuclear weapons and nuclear war are four main factors that still apply to many of us, that still and always must be combatted if we wish to remain responsible participants in our age.

• The first reason is that nuclear weapons and nuclear war belong to a field that is technically complex and rapidly changing. The public has, in general, been disposed to delegate difficult topics to people and agencies upon whom have been conferred the label of "expert."

• The second reason is that issues of national security have, since the end of World War II and the onset of the Cold War, acquired widely accepted immunity from lay analysis and critique. The world's democracies have tolerated much official secrecy and duplicity in the service of anti-communism and bureaucratic self-preservation.

• The third reason is that the subject of nuclear weapons and nuclear war carries its own psychological deterrence—it is sufficiently grim and overwhelming to dampen the interest of most of us, who wish to commit our minds to truth and beauty, our lives to health and happiness, and our salvation to the fruits of diligence.

• The fourth reason is that major estates and guilds in our societies have developed powerful, overlapping interests in pursuing a nuclear weapons policy that, although changing profiles with different governments has, since the start of the nuclear age, taken the same trajectory—an escalation in numbers of weapons, in systems complexity, and in strategic offensive and defensive doctrines. It has been hard to challenge the official national consensus of the Western nations, particularly when those in charge of the scientific, economic, and political communities all participate in and benefit from the current order.

The Turning Point

In 1980, the war in Afghanistan and the intensification of diplomatic hostilities between the U.S. and the U.S.S.R. shocked the U.S. public. Overseas tensions and militaristic rhetoric from many quarters provoked a tidal change in public outcry and consciousness throughout the world.

The barriers against becoming informed about the threat of nuclear weapons began to break down against the force of public concern. In the U.S., joined by many scientists and physicians, and by a few retired military and political leaders, the public began to close the gap in the debate about nuclear weapons and nuclear war, the gap between what is actually happening and what is publicly known.

As Dr. Schweitzer said so trenchantly in one of his radio broadcasts in 1958, it is not the physicists who must teach us about the effects of radiation on the world's biosphere. The responsibility lies with the biologist and the physicians.[8] In fact, we do need the physicist to help us understand the horrible dimensions of nuclear weapons and nuclear war. Oppenheimer and some of his colleagues who outlived him have been essential to our campaign of public understanding. But Dr. Schweitzer was correct in anticipating the effect that physicians would have on educating the public mind regarding the human and ecological consequences of such a catastrophe.

The U.S. Physicians for Social Responsibility (PSR) was founded in 1961, in the first wave of public reaction to nuclear testing. In the groundswell of public anxiety at the end of the 1970's and early 1980's, creating the second wave in public concern about nuclear weapons and nuclear war, PSR found new energy in its remarkable leader, Dr. Helen Caldicott. And in the early 1980's, the international movement of physicians, International Physicians for the Prevention of Nuclear War (IPPNW), was formed by Drs. Bernard Lown and Evgeni Chazov. We now have thousands of members worldwide.

New Insights

From the perspective of a physician who has participated in the debate for the public mind, Dr. Schweitzer early on identified awakening the force of public opinion as the most crucial bulwark against government excess.[9] Let me list a few milestones we have achieved in public consciousness and a few breakthroughs in understanding we have accomplished:

• *The futility of a medical response to a major nuclear war*: The scenario, changing with the weapons, has been accurately described to audiences since 1962. It is beginning to sink in.

We speak here of hundreds of millions dead outright in a global nuclear war involving two superpowers, these millions killed from the immediate effects of burn, blast and radiation. Fires would sweep the land, institutions would be leveled. Microestimates developed for the Western countries yield one surviving hospital bed for 100 to 500 seriously injured, one intensive care burn bed for each 10,000 such patients in need. There is no concerted or effective medical response possible in such circumstances.

• *The nuclear winter hypothesis* was developed by U.S. and U.S.S.R. scientists in 1983 and is now established as a reasonable predictor of probable outcome. We speak here of sudden drops in temperature ranging from one to 10 degrees centigrade, lasting for perhaps two weeks to three months, obliterating harvests, and inflicting great hardship in terms of exposure to cold and darkness on all those immediate survivors of burn, blast and radiation.

• *The illusion of civil defense*, an assessment precipitated by widespread public scrutiny of the U.S. civil defense plan but borne out by scrutiny of other countries' civil defense plans as well.

• *The Third World consequences of a major nuclear war in the Northern Hemisphere*, elaborated in 1986 and 1987. In the most trenchant example of this elaboration, the author of the SCOPE study said that, at the end of a nuclear war on this world, the world would look less like Hiroshima than Ethiopia.[10]

New Concepts

As a result of this work, and the work of many others, what do we now understand about the nature of nuclear weapons, the nature of nuclear war, and the facets of humanness that require us to pay attention to these facts? To begin with, nuclear war, once begun, whether by accident or intent, would not remain "limited."

Second, nuclear war cannot be described solely in terms of short-term effects. Nuclear war inflicts thorough and extensive devastation on all aspects of world existence over a very long time frame.

Third, nuclear war cannot be understood in conventional terms. It is not just a disaster greater than those we have witnessed, a disaster that brings with it standard opportunities for mitigation and response. This disaster, in its instantaneousness and totality, wipes out these premises and obviates response and recovery.

Fourth, and finally, a future war that involves the use of nuclear weapons can never be perceived as the event we have called "war" in historical terms. Wars, in the past, have been fought with the rational objective of winning. Listen to Winston Churchill, in May of 1940: "Victory at all costs, victory in spite of all terror, victory however long and hard the road may be; for without victory there is no survival."[11] That was in the pre-atomic era.

A future war in which nuclear weapons are used defines a scale of magnitude vastly and qualitatively different from weapons used in previous wars. A total of three megatons was exploded in the Second World War; in the Vietnam War, a total of four. We are discussing here potential yet postulated wars inflicting 5,000 to 10,000 megatons on each side. These weapons, because of the scale change they dictate, require a reevaluation of our interaction with our social and natural ecosystems— systems that we are discovering have aspects that are fragile as well as stable, mutable as well as enduring, here today and not necessarily here tomorrow.

These are major insights, building on enhanced understanding of the technology and its consequences that most people throughout the world are beginning to share.

The Choices Ahead

Significant strides have been made in progress toward arms control, if not toward Oppenheimer's dream of making future wars impossible. That we can say with confidence, despite the present time, as we look at the world in 1990 compared to what it appeared to be in 1980. There has been a sea change in consciousness and one objective change, the Intermediate Nuclear Forces (INF) Treaty signed in 1987— a small step, but a real one.

The events of 1989 have shaken loose political structures and unleashed popular yearnings in ways unimaginable to any observers even one short year ago. August 1989 was after Tiananmen Square and the killings in Tbilisi, but before the collapse of the Berlin Wall, before the revolutionary changes in Eastern Europe, before the accelerating events in the U.S.S.R. and South Africa.

And now, as the clouds of the Cold War lift, the light shines harshly on regional conflicts and newly armed contenders the superpowers helped create.

Facing all of us in the next decade will be many choices relating to peace and war, a further pursuit of the arms race or further reductions in weapons systems, a range of possible treaties relating to nuclear testing, nuclear proliferation, chemical and biological weapons, and conventional arms. Of particular importance and urgency are the Comprehensive Test-Ban (CTB) Treaty and the Chemical Weapons Convention. Both of these are very near conclusion, both require positive U.S. action to succeed.

The fruits of our failure to act so far stare up at us from the headlines on the front page.

Embedded in the choices we make, the agreement we dare to negotiate or dare to discard, will be decisions we make, consciously or not, about the kind of world we wish to become and the kind of security we will strive to achieve. It is past time to frame our view of the world in a context we have far too long ignored—global interdependence.

Some essential elements in a political recasting of this new sense of security include: defining our vital interests within the constraints of interdependence; assessing the enemy threat with more maturity and less rhetoric; recognizing the autonomous local nature of many regional conflicts; sharing the burdens and responsibilities of power and wealth with our allies who are becoming equally weighty participants in the world community; and affirming the extent to which true security lies in the economic and moral integrity of nations working together.

Conclusion

In the past three years, the U.S. Physicians for Social Responsibility spent some time drafting a mission statement, in recognition of our impending passage into the 21st century. Nuclear weapons are but one technology, now accessible; there are many on the horizon to which a rapacious, trapped, or brutal nation could turn. We define for ourselves the elements that brought us together and the goals for which we would strive. Because these sentiments are common to many people, working in many different professions and ways of life, I would like to share part of this mission statement with you. As I read the paragraphs in the presence of the man we honor today, I am sure you will hear some echoes of his voice.

Our values

As a result of training and choice, we believe:
- That human life is precious, powerful, but vulnerable;
- That human life draws sustenance and coherence from the biological and

social systems in which it is enmeshed;

• That knowledge results from experience, reconsidered through discourse and exchange, tested against new experience, and evaluated in settings open to public review;

• That the use of knowledge imposes responsibility to protect the quality of life, not destroy it.

Our objectives

Impelled and guided by these values, we work:

• To eliminate the use of weapons of mass destruction;

• To establish conditions for constructive and peaceful ways to settle or contain national and international conflict;

• To explore the psychological, social and biological roots of our individual and group behavior in order to develop ways to live our lives and raise the generations that follow in paths that lead toward life and continuity.

• To promote the physical and psychological well-being of humanity.[12]

These are statements of belief and commitment, of dreams and responsibilities. If we as a profession, as a nation, and as a world are to enter the 21st century under full sail, we must look back to gain courage and insight; we must look forward to acquire some sense of direction; and we must look within. If it can become our common dream to make future wars impossible, it must first become our common responsibility.

[1] George Marshall in commencement address at Harvard, June 5, 1947, as quoted in Isaacson, W. and Thomas, E. *The Wise Men*. New York: Simon and Schuster, 1986: 413.

[2] Isaacson, W and Thomas, E. *The Wise Men*. New York: Simon and Schuster, 1986: 314.

[3] Woodrow Wilson, address to Congress, April 2, 1917, in Hofstadter, R. and Hofstadter, B.K., eds. *Great Issues in American History: From Reconstruction to the Present Day, 1864-1981*. New York: Random House, 1982: 208.

[4] Interim Committee Scientific Panel, quoted in Rhodes, R. *The Making of the Atomic Bomb*. New York: Simon and Schuster, 1986: 751-52.

[5] Oughterson, A.W. and Warren, S. *Medical Effects of the Atomic Bomb in Japan*. New York: McGraw-Hill Book Co., 1956.

[6] Jack, H.A., ed. Albert Schweitzer: On Nuclear War and Peace. Elgin, IL: Brethren Press, 1988: 69-91.

[7] Center for Defense Information. The Defense Monitor, 1988; 17: 5.

[8] Jack, H.A., ed. Albert Schweitzer, 1988: 74.

[9] Jack, H.A., ed. Albert Schweitzer, 1988: 67, 84, 135.

[10] Harwell, M.A., Harwell, C.C. "Nuclear Famine: The Indirect Effects of Nuclear War." In F. Solomon, R.Q. Marston, eds., The Medical Implications of Nuclear War.

[11] Churchill, W. Parliamentary Debates. House of Commons Official Report. Fifth Session of the 37th Parliament of the United Kingdom of Great Britain and Northern Ireland. 4 George VI. 360 H.C. Deb. 5S. London: His Majesty's Stationery Office, 1940: 1502.

[12] Physicians for Social Responsibility. Mission Statement. PSR Reports. Washington, D.C. Summer 1988: 5.

The Role of the Individual in Achieving World Peace

By Norman Cousins

If the peace of the world depends on governments, there will be no peace. Governments, especially the major governments, are the *problem*, not the answer. Peace depends on the extent to which national sovereign states can be integrated into a world system of peace. And this will be possible only if individuals prod their governments into an adequate world organization.

Today, as we consider the control of arms—especially nuclear arms and other weapons of mass destruction—we think in the context of the Iraq crisis. And I think it is natural, since we meet in the name of Albert Schweitzer, to ask: what would Schweitzer say?

There is a tendency, and certainly a temptation, to project on Albert Schweitzer our own view of current events. I have tried to imagine him sitting on our panel and it seems to me that he would be proud that the United Nations exists, proud of the Security Council's condemnation of Iraq's invasion of Kuwait and of its efforts to peacefully resolve this crisis.*

And yet, he would ask why it was necessary for the world to wait until after a crisis occurred. He would ask what is being done to prevent future such crises. He would ask about the persistent refusal of the larger nations to build the United Nations into an organization capable of dealing with the basic causes of war.

* The International Albert Schweitzer Colloquium was held within a few weeks of Iraq's August 1990 invasion of Kuwait and prior to both the United Nations approval of military action against Iraq and the commencement of that action.

Yes, Schweitzer would take pride, as a human being, in the existence of the United Nations, but he would also issue a challenge to the world to finish the job begun at San Francisco in 1945: to develop the United Nations into an organization capable of enacting and enforcing world law.

And while he might praise President Bush for going to the Security Council, I think Schweitzer would wonder why he would then engage in an exchange of insults and threats with Saddam. I think he would wonder why, instead, President Bush would not be going to the World Court—because this is what the World Court is for. But Schweitzer would also recognize that the moral position of the United States going to the World Court has been weakened by its refusal to accept the jurisdiction of the Court in Nicaragua.

And so, Schweitzer would look back at all the past mistakes, the failures, the arrogance of the national sovereign states, and he would call upon individuals all over the world to recognize that the peace of the world depends on *their* ability to insist that the national governments at the United Nations get on to the main business before them: institution-building—creating the United Nations into an organization that does not run after history with a mop.

With respect to the specific problems of arms control and the nuclear threat, I think Albert Schweitzer would probably say that we have to connect these concerns to the problem of creating a durable peace. For example, he might point out that as the United States and the Soviet Union agree between themselves to cut back on arms, the Soviet Union is in need of hard currency and is seeking to create funds for itself by dumping its arms on the world market. This is a very grave threat.

Schweitzer would remind us that the United States has also engaged in the world arms trade. He would want the United Nations to make sure that arms reduction does not result in the indiscriminate dumping of arms, especially in the Third World. He might also call attention to the role of the United States in encouraging Iraq at the very start of its war with Iran, and to the arms supplied to Iraq by both the United States and the Soviet Union. In short, I think that Schweitzer would suggest that we all ought to wash our hands if we are to take adequate hold in grappling with the questions of peace.

We need to ask ourselves about the moral implications of Schweitzer in the modern world: what would he represent to us today? I think the answer is probably to be found in your presence here, and in your faces, because it seems to me that Schweitzer's greatest power is represented by his ability to bring out the best in us and to make us hungry to apply that best. We are here not just to honor Schweitzer, but to ask ourselves what we can most usefully do in the present situation. We take nourishment and inspiration from his memory.

We recognize that his contribution to peace has been not so much to our understanding of the principles of peace, as to the recognition of our responsibility to build as we have to build, to prod as we have to prod, in the cause of peace.

When I think back on Schweitzer at Lambarene, I think of his prodigious, very practical labors. Often he worked around the clock. I especially remember his great good humor under those circumstances, and the many amusing stories he loved to tell that would brighten the day for all those around him. He urged us never to lose our sense of humor; he regarded humor as a regenerating force in life, which it is. And he urged us never to lose confidence in ourselves.

Albert Schweitzer would tell us that although a problem might be great, our ability to meet the problem is even greater. It seems to me that this is a useful message for us at a time when everything looks bleak, at a time when we feel there is no place to take hold. Schweitzer would say to us: there is a place to take hold. We must recognize that the individual in today's world is sovereign, that the individual holds the ultimate power, that we have the obligation to move nations as they have to be moved and we are in exactly the right place to begin.

Albert Schweitzer as Senior Citizen:
Dealing with Nuclear War and with Colonialism

By Homer A. Jack

Albert Schweitzer was born on January 14, 1875 in Alsace, then Germany, and he died on September 4, 1965 in the independent African state of the Gabon. In many cultures, he would have been considered a senior citizen from 1940—age 65—until his death, although as far as I know he never received a monthly pensioner's check or social security check either from French Equatorial Africa or the Gabon.

After the Second World War, Albert Schweitzer in the last 20 years of his life was increasingly a well-known figure. He was an inevitable target for all kinds of international hucksters in the 1950's and 1960's. Dozens of people made their way to Lambarene, invited and uninvited, in the hope that Schweitzer might buy their particular idea or program. Each insisted that if Schweitzer only endorsed his or her project or product, the world would be a much better place in which to live. Few of these ardent visitors ever made a "sale" at Lambarene. Most quickly found themselves on a canoe, leaving Lambarene, and only belatedly realizing that Schweitzer had sold his philosophy of Reverence for Life much more than he bought any panacea or nostrum from them.

However, two global, ethical issues in the 1950's were presented to Albert Schweitzer that he could not summarily dismiss, especially after he received the Nobel Peace Prize. One issue was the hope that Schweitzer might speak out against nuclear weapons testing and nuclear deterrence. The second was that he might speak out against continuing colonialism in Africa. Both requests came to Schweitzer when he was a *senior* senior citizen, past his mid-seventies.

I was among the many visitors to Lambarene, for the first time in May of 1952.

I did not come to sell Schweitzer anything. On a two-week visit, I simply wanted to know Schweitzer, to find out about his life and thought. On the other hand, toward the end of my visit, I could not help but ask him leading questions about the then-new United Nations and about disarmament and colonialism.

My visit was satisfying in the extreme. The only disappointment to me, as an unabashed American liberal, was that I found Schweitzer conservative and even what I considered "old-fashioned" on both of these important issues. Yet who was I to suggest that Schweitzer should publicly ask for nuclear disarmament or for African freedom? My journal records that I occasionally, and very hesitantly, differed with him on these issues of disarmament and colonialism. Yet I, then a 36-year-old American clergyman, knew that my place, my role, was hardly to ask this senior citizen to question, let alone change, his firmly held views.

That was in 1952. The next year the Norwegian Nobel Committee announced that Albert Schweitzer was the Nobel Peace Laureate for 1952. He could not immediately travel to Oslo, and so he was given the award in absentia in December 1953. At that time, Gunnar Jahn, chairman of the Nobel Committee in Oslo, declared that Schweitzer "has shown us that a man's life and his dream can become one." He added that Schweitzer's "work has made the concept of brotherhood a living one, and his words have reached and taken root in the minds of countless men."[1] The jungle physician was awarded the prize for who he was and how he inspired others with his humanitarian compassion, not because of any specific activity within a narrow definition of peace. Jahn's judgment about Schweitzer was similar to that articulated by Norman Cousins: "It is not so much what Schweitzer has done for others, but what others have done because of him and the power of his example."[2]

On March 1, 1954, an American hydrogen bomb test over the Bikini Atoll in the South Pacific resulted in radioactive "ashes of death" falling on Japanese fisherman. Within a month, Indian Prime Minister Jawaharlal Nehru urged a moratorium on all nuclear tests. Pressure on Schweitzer was building for him to talk publicly on this and other issues, especially since he was now a Nobel Laureate. He was reluctant to become an instant pundit, but his peers urged him to speak. One was Albert Einstein.

The two Alberts—Einstein and Schweitzer—were born in Germany only 125 miles and four years apart.[3] They corresponded about nuclear issues early in 1955, but then Einstein died in April of that year. Also in 1955, U.N. Secretary-General Dag Hammarskjold asked Schweitzer to "send forth an essential message to the world" and "animate international life with a new spirit."[4] Schweitzer also corresponded, in this period, on nuclear arms issues with Bertrand Russell, Linus Pauling, Adlai Stevenson and Pablo Casals. But it was clearly Norman Cousins whose visit to Lambarene in January 1957 made Schweitzer a different person.

This visit is described in some detail by Cousins in two volumes, the second giving the details of what can only be called the conversion of 82-year-old Schweitzer by 42-year-old Cousins.[5] Cousins was indeed a contemporary American missionary to Africa, bringing a religious if non-sectarian urgency to act for nuclear peace. In May of 1957, Schweitzer wrote to Cousins, "Without your coming here I don't think I should have decided to make the statement..."—the "statement" being Schweitzer's worldwide radio broadcast entitled *Declaration of Conscience*. Schweitzer added in his letter to Cousins, "You were right to encourage me to do it. I will never forget that."[6]

I must emphasize that Schweitzer was 82 years of age when Cousins arrived. At this age, when most human beings rest on their laurels, even their Nobel laurels, Albert Schweitzer began the study of nuclear fallout and possible nuclear war. He read, he corresponded, he welcomed scientists and politicians to his jungle study. Thus it was, three months after Cousins's visit, that Schweitzer composed his *Declaration of Conscience*, concluding that "the end of further experiments with atomic bombs would be like the early sunrays of hope which suffering humanity is hoping for." This declaration was broadcast over Radio Oslo, and to much of the world, on April 23, 1957. It was completely out of character from anything Schweitzer had previously done. Now he actively sought publicity for his views, after spending a decade or more trying to avoid publicity of any kind. Here we have a senior citizen in his ninth decade studying a very complex subject, eager to inform and to reform the world. A new Albert Schweitzer clearly emerged.[7]

Africa and colonialism was a second global, ethical issue about which some visitors to Lambarene also tried to change Schweitzer's mind. Schweitzer wrote in his *Memoirs of Childhood and Youth*[8] of being taken to nearby Colmar and seeing a "stone figure of a Negro," from the chisel of Frederic-Auguste Bartholdi, an Alsatian who was also the sculptor of the Statue of Liberty. Schweitzer wrote that "the face (of the Negro) wears an expression of thoughtful sadness which I could not forget... The countenance spoke to me of the misery of the Dark Continent."

That experience was in the 1880's. In 1913 Albert Schweitzer and his wife first went to Africa. Despite two world wars, he spent much of his life there. In the mid-1940's, just when Schweitzer technically became a senior citizen, an historic correspondence occurred on the subject of Africa. The black American scholar and activist, W.E.B. DuBois, himself age 78, contributed a chapter to a 70th birthday volume, a *Festscrift*, honoring Schweitzer.[9] DuBois's essay was only mildly critical of Schweitzer. DuBois insisted that Schweitzer had "no broad grasp of what modern exploitation means, of what imperial colonialism has done to the world. If he had, he probably would have tried to heal the souls of white Europe rather than the bodies of black Africa." Then DuBois was charitable: "Since, however, [Schweitzer] did not comprehend this, and since nothing in his surroundings brought him any such clear apprehension, he deserves every tribute that we can give him for trying to do his mite, his little pitiful mite, which in a sense was but a passing gesture, but which perhaps in the long run will light that fire in Africa that will cleanse the continent and the world." DuBois concluded that Schweitzer in Africa "was neither deceived nor enlightened" about colonialism. Rather, Schweitzer "saw the pain and degradation of this bit of God's earth as something he could alleviate; for many years he gave his life to it."

A decade after this Schweitzer-DuBois correspondence, I paid my second visit to Schweitzer at Lambarene, in March 1957. I had just come from observing the freedom celebrations when the Gold Coast became Ghana. I brought with me the elan, the spirit of exhiliration of watching the moment of independence of the first African state after World War II. I described how many blacks from around the world were in Accra for that event, including Martin Luther King, Jr. However, my joy was not well received, neither by Schweitzer, nor by the nurses and physicians around him. My purpose was again to visit, not to propagandize. Indeed, I did not know at the time that Norman Cousins had been there barely two months earlier, on his nuclear mission. But on this, my second visit, Schweitzer was now a Nobel Peace Laureate and it was relevant for me at least to suggest that French Equatorial Africa might one day go the way of the Gold Coast. Schweitzer heard me out, while fingering souvenirs of the Ghana independence ceremonies which I gave him. Even in my first visit, Emma Hausknecht, Schweitzer's veteran nurse, volunteered, "Poor Dr. Jack, he loves the Africans." Schweitzer sharply interrupted, "So do I, but I know their weaknesses, too."[10]

A year later, 1958, I was back at Lambarene, this time with more pamphlets and clippings about African freedom, since I had just come from being an observer at the All African People's Conference, also at Accra, attended by Patrice Lumumba, Julius Nyerere, and Tom Mboya, with Prime Minister Kwame Nkrumah as host. Again, Schweitzer was polite, very curious, even probing, but his skepticism was obvious. He did not change his outlook on this issue.

Schweitzer once told me that "the United Nations gives Africans ideas of freedom and self-government . . . [but] the goal must be freedom from hunger before independence." He observed that the "Africans should first be well fed. Get them to work so they can eat." Schweitzer added that I came to Lambarene to hear theological heresies, and now I was hearing other kinds of heresies. He declared that "if one were to define Albert Schweitzer, let them say he is a bundle of heresies brought together with robust thinking."

Why, in his eighties, did Schweitzer change his mind about not speaking out and soon relish action on a world scale against nuclear tests? Why did Schweitzer become possessed in trying to stop nuclear tests? Given his Nobel Peace Prize, I believe Schweitzer wanted to take leadership on an important peace issue. Yet why did he also hold on to his deep skepticism about the winds of change in Africa? His experience in Lambarene over several decades simply did not lead him to advocate immediate African freedom. Also, perhaps the Gabon, or at least Lambarene, was slower to welcome the changes that were occurring in Accra, in Nairobi or in Leopoldville.

Most important, perhaps few persons came to Lambarene, in the fashion of Norman Cousins, to attempt to convert Schweitzer on the colonial issue. Several visitors may have tried, but they were not persistent, certainly not as successful as Cousins with nuclear testing. My own fleeting "suggestions" about African freedom during four visits Schweitzer could, and did, promptly dismiss.

The relationship of Schweitzer to Africa and colonialism is not as simple as I may have portrayed. It has many layers and is being carefully studied over several years by a working group related to the Albert Schweitzer Fellowship here in the United States. As for DuBois and Schweitzer, both died in Africa and are buried there. DuBois, age 95, died August 27, 1963 at Accra, on the same day as Martin Luther

King's remembered March on Washington. And Schweitzer, age 90, died on September 4, 1965 at Lambarene.

What would Albert Schweitzer think if he were alive today, 25 years after his death? He would be amazed, as all of us are, at the rapid changes in Eastern Europe. Schweitzer was never seduced by communism, as were a few of his fellow European intellectuals, including his second cousin, Jean-Paul Sartre. (On a visit, I asked Schweitzer if Sartre were a relative. Schweitzer replied, "Yes, but neither of us admits it.")

While Schweitzer today would be very pleased at the end of the Cold War, he would be exasperated that underground nuclear weapons tests have still not been outlawed by treaty. He would be sad to know the statistics: a total of more than 1,100 tests have been detonated since his death.[11] These tests were mostly underground, except for 66 in the atmosphere by China and by his own France. This absence of progress in nuclear disarmament might have induced Schweitzer, if alive today, to want to take to the radio again with another *Declaration of Conscience*, this time insisting that the United States and England join the 116 other states party to the existing Partial Test-Ban Treaty and broaden it to include a ban on all underground nuclear tests. Yet Schweitzer, were he alive, would now realize that, thanks primarily to Mr. Gorbachev, the threat of nuclear war has greatly diminished in the past two years despite the lack of deep cuts in the nuclear arsenals of the super powers.

As for Africa, I hesitate to suggest what Albert Schweitzer would now think. In some ways, he would have found many reasons, given events in Africa in the 1970's and 1980's, to indicate that he was right after all about Africa. Yet his own character would surely make Schweitzer shrink back from any personal hubris. He would probably shake his head in disbelief to know that even his hospital has been modernized and that great changes have come to the Gabon, now a member of that erstwhile rich man's club called O.P.E.C.

Schweitzer would shyly listen to accounts over the past quarter of a century on the work of the organizations bearing his name, in several parts of the world, that are trying to spread his multiple legacies. He would probably shrug his shoulders when told about this international Colloquium at United Nations headquarters. He might suggest, instead, that his friends use this occasion and this conference room to demand a treaty to end underground nuclear weapons testing. As for the current United Nations African agenda, Schweitzer might take off his white pith helmet, rub his powerful fingers through his wet hair, and become "as mute as a carp."[12]

---

[1] *Nobel Lectures. Peace. 1951-1970.* Volume 3. Edited by Frederick W. Heverman. Amsterdam: Elsevier. 1972: 37-45.

[2] Cousins, Norman. *Dr. Schweitzer of Lambarene.* New York: Harper and Row, 1960: 215.

[3] Jack, Homer A. "The Same First Name." *The Courier* (of the Albert Schweitzer Fellowship). Summer 1986: 5-9.

[4] Brabazon, James. *Albert Schweitzer: A Biography*. New York: Putnam, 1975: 429-30.

[5] Cousins, Norman. *Op. Cit.* See also *Albert Schweitzer's Mission: Healing and Peace*. New York: W.W. Norton, 1985.

[6] Cousins, Norman. 1960, *Op. Cit.*: 190.

[7] Jack, Homer A., Editor. *Albert Schweitzer on Nuclear War and Peace*. Elgin, IL: Brethren Press, 1988.

[8] Schweitzer, Albert. *Memoirs of Childhood and Youth*. New York: Macmillan. 1925: 63.

[9] DuBois, W.E.B. "The Black Man and Albert Schweitzer," *The Albert Schweitzer Jubilee Book*, edited by A.A. Roback. Cambridge: Sci-Art Publishers, 1945: 121-27.

[10] This material may be found in the Swarthmore College Peace Collection, Swarthmore, PA, DG63, Series VI, Box 9, African Diary: 19.

[11] *SIPRI Yearbook 1990: World Armaments and Disarmament*. New York: Oxford University Press, 1990: 56-57.

[12] Schweitzer, Albert: *Out of My Life and Thought*. Translated by A. B. Lemke. New York: Henry Holt, 1990: 115.

We Must Work to End War

By Linus Pauling

My wife and I had one of the great experiences of our lives when we spent two weeks with Dr. Schweitzer in Lambarene. Every evening after dinner he asked me to come to his room for an hour. I don't remember that I ever said anything; I listened to what he had to say. He was twenty-six years older than I, an older generation in my eyes, and a sort of father figure.

During our visit we also became acquainted with some of the remarkable people—volunteers, for the most part—who were there with him. We came away from Lambarene with an increased feeling of respect and admiration for Schweitzer.

But we knew another Albert better than Albert Schweitzer; that was Albert Einstein. I was, for many years, a member of the board of trustees of the so-called Einstein Committee, the emergency committee of atomic sciences that worked to educate people about the nature of nuclear weapons and nuclear war.

Einstein, like Albert Schweitzer, was a great worker for world peace and for human rights. As early as 1946, Einstein said, "Now that a single bomb can destroy a whole city, there must be a change in our thinking. We must abandon war as the mechanism for settling disputes between nations and accept international law in its place."

He was talking then about the fission bombs such as had been used on Hiroshima and Nagasaki—each one equivalent to approximately 20,000 tons of TNT in its explosive power and such that a single bomb could, and did, kill 100,000 or 150,000 people and destroy a city.

Only a few years later, super bombs had been developed with a thousand times greater destructive power per weapon. The twenty megaton bombs, each equivalent to 20 million tons of TNT, were such that one of these bombs, exploded over New

York City, could kill 10 million people and destroy the city. As the United States and the Soviet Union—and, later, other countries—constructed these astonishing weapons, it became evident to people who studied the nature of nuclear war that a war between the United States and the Soviet Union could mean the destruction of civilization and the extinction of the human race.

So Einstein's 1946 statement became more significant than ever. The existence of tremendous stockpiles of nuclear weapons makes it unthinkable that there should be a war between the United States and the Soviet Union. One might say: Well, why don't they agree to use only conventional weapons? My answer is: Why don't they agree not to engage in war?

For many years I have said that war between the great powers, the nuclear powers, is ruled out by the existence of nuclear weapons stockpiles. It doesn't make much sense to be spending seven percent or more of the country's gross national product on weapons for a war that cannot be fought so long as there is any sanity left in the human race. I argued that treaties should be made, and some very significant treaties have been made.

Seven years ago I wrote a paper arguing that the time had come for the United States to begin taking unilateral action toward disarmament and demilitarization because (and I wasn't the first to contend this) such action—especially with regard to nuclear weapons—would be safer than continuing to build weapons stockpiles. There was a modest precedent for nations taking unilateral action: In 1959, the Soviet Union announced that it was stopping all testing of nuclear weapons unilaterally, and that this action would continue so long as no nuclear weapons were tested by the Western powers. This lasted until 1961 when France carried out a nuclear weapons test. Soon thereafter, both the Soviet Union and the United States resumed their nuclear weapons tests. The effort to move toward sanity failed.

Other unilateral actions, primarily by the United States, were not in the direction of disarmament. They involved developing new nuclear weapons and new methods of delivery. We usually led the Soviet Union in these developments by about four years.

But now we have seen the Soviet Union under President Gorbachev taking a great unilateral action, and this is not surprising. For decades the stated policy of the United States has been to develop military weapons at considerable burden to our economy, knowing that as long as the Soviet Union felt the need to keep pace, the burden would be *twice* as great for their economy. Recently I have read that our gross national product is more likely *three times* that of the Soviet Union. So if we spend seven percent of our gross national product on militarism, they're spending about 21%. It's not surprising, then, that the economic situation would become so serious as to essentially force Gorbachev to take action. And, of course, it would be good for us to take actions to improve the economy of the United States as well.

In addition, increased efforts should be made to make, sign and approve international treaties controlling weapons—treaties to reduce stockpiles, to prevent the spread of nuclear weapons to other nations, to stop the development of new weapons and means for delivering them, and to create a complete moratorium on nuclear weapons testing. Stockpiles of nuclear weapons are a continuing threat to the human race. There is always a chance that something will go wrong, that some accident will occur leading to a devastating series of actions and reactions and culminating in the end of our civilization.

I do not advocate nuclear disarmament in the sense of destroying all nuclear weapons. Rather, I believe these stockpiles should be reduced from their present completely insane levels to somewhat less irrational levels. We should then have these diminished stockpiles under controls, both national and international, until finally—how long, perhaps a hundred years from now—their existence would be such an anachronism that the effort would be made to find some safe way of disposing of the fissionable materials.

Together with our government, we should be working vigorously to get arms control treaties made, signed and put into effect. And with the money we save, just think what can be done toward solving our other great problems—the deterioration of the environment, the population problem, starvation and malnutrition throughout the world, the failure of the human race to have developed a system that permits each human being to have at least the possibility of leading a good life.

I have always been proud of being an American, born in Oregon, and I continue to be proud of my country. So I should like to see the United States—not Gorbachev and the Soviet Union, but the United States—taking the lead in the great steps that need to be made. These steps are obvious: a decrease in money wasted on militarism and the discovery of ways to solve other great problems in the world. One of these problems is that great nations have made money by providing modern weapons to other nations. I remember about thirty years ago an Assistant Secretary of State being given the State Department Award for his patriotism and zeal in having sold a billion dollars worth of weapons to the underdeveloped countries of the world. In his acceptance speech he said that with the proper effort it should be possible for the United States to be selling 15 billion dollars worth of these weapons within ten years.

My estimate is that the United States, the Soviet Union, Sweden, Great Britain, France and West Germany are now selling—or giving—100 trillion dollars worth of weapons to other countries. And now there is the danger that the United States and the Soviet Union will try to profit in the reduction of their weapons stockpiles by selling these weapons to other nations.

Great nations should not be taking the actions they have—of starting wars or supporting wars involving other nations. We should have cooperation between the United States, the Soviet Union and other leading nations in trying to settle disputes— some of which have been going on for a couple of thousand years—that lead to wars. I am confident that if this were a problem we really made an effort to solve, it could be solved. But it will not be solved by continuing to provide advanced military weapons to nations all over the world, either by sale or gift.

I like to solve problems, especially problems that are presented to us by nature. We have developed a reasonable understanding of nature, but there are still many thousands of aspects of nature that we don't understand. That's what I enjoy: solving—or trying to solve—scientific problems. I don't enjoy giving lectures about world peace, but I do it. Perhaps I was prodded into it by my wife who, in 1945, after the atomic bombs on Japan, thought that it was my duty to do what I could. I think everyone has the duty to do what he or she can to solve these problems: by writing letters to the editor, joining peace groups, demonstrating or whatever is possible for each individual to do.

Every once in a while I am asked, are you hopeful for the future? I have always been hopeful for the future; perhaps it's just part of my genotype, but I can also give an argument about it. I think that you and I are really fortunate to be alive at an extraordinary period in the history of the human race—the demarcation between the past millenia of war and suffering and a great future of peace, morality, justice and human well-being. We can see this as a possibility if the resources of the world are used in the right way and if there is not the population catastrophe that might prevent it.

So the first thing to work for is the abolition of war and the development of methods for preserving world peace. If we make proper use of the earth's resources, of discoveries made by scientists, of the efforts of mankind and the work of human beings, I believe we can then build a world characterized by economic, political and social justice for all human beings and a culture worthy of man's intelligence.

Medicine
and
Health Care

Miller

Foege

Sidel

Cirincione

Karefa-Smart

Jampolsky

Moderator's Introduction

By David C. Miller

Beginning in 1913 and over the following 52 years, in the hospital village he built near Lambarene in the equatorial African jungle, Dr. Albert Schweitzer provided medical and surgical care for tens of thousands of Africans living in that region.

Not long after his death in 1965, a new and modern Schweitzer Hospital was built alongside the old village hospital, and has continued to serve the people of Gabon.

These achievements, impressive as they are in their magnitude and dedication to the relief of suffering, have become subject to considerable criticism and controversy; some of it quite valid, some otherwise.

While we shall not have time today to deal with these questions in any depth, we trust that a more balanced and mature view of them may emerge from this afternoon's presentations and discussions by our distinguished panel of speakers.

We shall pay tribute to what Dr. Schweitzer did in his time, in his way, with his limited resources, in his little part of Africa. But our main intent is to grapple honestly and realistically with today's and tomorrow's urgent problems in health care and medical care, both in the poorest nations of Africa, Asia, and Central and South America, and in our affluent part of the world.

And we may well find that Albert Schweitzer, for all the shortcomings in his valiant efforts for healing in Africa, was, after all, a pioneer in more ways than one, a man whose remarkable spirit and example can still guide and inspire us today in our efforts to bring better health to the world.

To Give is to Receive

By Gerald G. Jampolsky and Diane Cirincione

Editor's Note: This presentation was made alternately, as indicated, by Dr. Jampolsky and Ms. Cirincione.

Jampolsky: There is a way to live in this world with compassion and full commitment each second that we are here, and to hold with love all that is living and treat it with tender loving care.

The world is in dire need of models that exemplify these principles, and Dr. Albert Schweitzer continues to shine as a model of what each of us can do when we take a leap in faith and trust and live a life, directed by our inner voice, toward helping others.

Cirincione: We are humbled and honored to be here today to take part in this contemporary celebration of Dr. Schweitzer's life. What inspires both of us the most about the life of Albert Schweitzer is that he *lived* the principles of love, forgiveness and reverence for all of life, instead of just preaching these principles to others.

Here was a man who was able to detach himself from the daily distractions of life that so many of us think make us happy. He knew that a life of helping and a life of prayer were one and the same; that helping is an active life of prayer. He did not pretend to be perfect, but he did his best to live his life with honesty and integrity, and to create a harmony in what he thought, what he said and what he did. This, to us, exemplifies a truly honest life.

As we celebrate Albert Schweitzer, it is not only important to celebrate his spirituality and his accomplishments, but also to acknowledge and respect his humanness and his own personal struggles. Like the rest of us, he had to struggle with his ego and with his own problems of judgment, irritation and anger.

The fact that he was in no way perfect gives all of us a tremendous amount of hope. Why? Because his humanness and his own struggles help us to bridge the gap between Schweitzer and ourselves. And because we can relate to the humanity in him, we receive the greatest gift you can get from one who has been here on earth before you—inspiration. When we read or hear about his imperfections, we can learn to identify with his light (as Jerry [Jampolsky] has said for so many years), and not his lamp shade, not his costume.

By learning from him, and not keeping him on a pedestal, we can more easily recognize that he was, in some ways, an ordinary person who did extraordinary things. And if I think of him in his ordinariness, then I know that perhaps I too can do something extraordinary in this world. But if I think he is only extraordinary, then I want to give up. So I would rather relate to him as a very *human* being. He did not always spiral upward on his spiritual journey, but he did make a consistent and determined effort to do good in the world. He taught the world that when one person can sincerely ask how he or she can be truly helpful, that person can have an enormous and lasting impact on the world.

Jampolsky: What this can mean to each of us is that, as ordinary people, we too can do extraordinary things when we are totally committed, have total and complete faith in God, and are convinced that love is the answer, no matter what the problem, no matter what the question.

On the other hand, many of us have egos that make us quick to compare ourselves with others, and when we do that we develop a "cancer of the soul." Let us look instead at the examples that Dr. Schweitzer gave us throughout his life of what positive things can happen—in situations that look impossible—when we step aside and let God lead the way.

As we celebrate and learn from Dr. Schweitzer, may we remind ourselves of one of the most powerful statements of Jesus, "What I can do, you can do, and more." Dr. Schweitzer's life so beautifully demonstrates that when we commit our lives to be helpful to others, the presence of God's peace is felt by all.

I think his life also demonstrates to us that one person not only can but does make a difference. I think that Mother Teresa probably sums it up better than anyone when she says, " It is not how much we do in life that counts, but how much love we do it with."

So when we think that Albert Schweitzer should have been out in the world more publicly, it's perhaps because we're judging wrongly. Perhaps quietly working with the villagers was the most powerful thing he could have been doing, because the power of that love has so obviously spread.

To us, Albert Schweitzer's life is like connective tissue that brings into harmony religion, philosophy, medicine and music. The symphony of his life was not always melodic, yet its direction was always clear. We are most fond of the following quotes which represent his way of living.

On attitudes: "The greatest discovery of any generation is that human beings can alter their lives by altering the attitudes of their mind."

Cirincione: On ethics: "Let me give you a definition of ethics: it is good to maintain life and further life; it is bad to damage and destroy life. However much it struggles against it, ethics arrives at the religion of Jesus. It must recognize that it can discover no other relationship to other beings as full of sense as the relationship of love. Ethics is the maintaining of life at the highest point of development—my own life and other life—by devoting myself to it in help and love, and both these things are connected."

Jampolsky: On responsibility: "I have a responsibility for all the good that has happened in my life and a sense of duty to pay for it by helping others."

Cirincione: On sincerity: "Not less strong than the will to truth must be the will to sincerity. Only an age which can show the courage of sincerity can possess truth which works as a special spiritual force within it. Sincerity is the foundation of the spiritual life."

Jampolsky: On the way to peace and away from violence: "As one who tried to remain youthful in his thinking and feeling, I have struggled against the facts and experience on behalf of belief in the good and the true. At the present, when violence clothed in life dominates the world more cruelly than it ever has before, I still remain convinced that truth, love, peaceableness, meekness and kindness are the power which can conquer all violence."

Cirincione: On Reverence for Life: "I have tried to relate Christianity to the sacredness of all life. Why limit Reverence for Life to the human form?"

Jampolosky: Yet in spite of these ancient truths that are known only too well by all of us, there is an ego voice, a "doubting Thomas," that challenges us with the question, "Can what I do really make a difference?" It is something that we struggle with weekly, daily, in times when we forget to throw away our measuring sticks.

Cirincione: A dear friend of ours, Wally "Famous" Amos, of chocolate chip cookies fame, shared with us a story that shifted our perception about the value of what one person can do. This makes me think of Dr. Schweitzer in Africa.

The story goes that a man was walking on the beach on the windward side of Oahu, in Hawaii. There had been a tremendous storm and, from that storm, thousands and thousands of starfish had been washed up on the beach. It was very clear that, by the next tide, the starfish were not going to get back in the water; they were all going to die. The man walked down the beach, depressed, thinking, "This is so sad. This is such a hopeless situation."

He kept walking, feeling worse and worse, until he came upon a little old woman. She was kneeling on the beach, throwing starfish back in the water one by one. And the man said to her, "Isn't this the most tragic situation you've ever seen; it's so sad!

What are you doing? What possible difference can this make? There are tens of thousands of them and they're all going to die."

The woman leaned down, picked up a single starfish, gently threw it in the water, and said, "It made a difference to that one."

So we'd like to think that every act of goodness, no matter how small or seemingly insignificant, is like throwing one starfish back into the sea.

Jampolsky: We are learning at our Centers for Attitudinal Healing that the most precious gift we can give to another person is our inner peace and our unconditional love, a love that defies any kind of measurement.

Most likely each of us, during the course of life, asks, "What is the purpose of life?" Unfortunately, we often end up thinking only of ourselves and of our immediate families. We believe that the world is screaming for us to awaken to the truth of our own innocence and the innocence of others. "And a little child shall lead them . . . "

Cirincione: We've been blessed to work with children over the years. They serve as wonderful teachers to us, particularly when we look upon them as equals. Some time ago, we had a wonderful experience when we were in Aspen, Colorado, staying at a friend's home. Jerry and I were quietly sleeping outside on lounge chairs. When we awoke, there was a beautiful child there, a six-year-old named Jessica. We found out that she lived down the hill with her mother.

After awhile, Jerry asked one of his typical Jampolsky questions that I never would have thought to ask this little girl. "Jessica," he said, "what do you think the purpose of life is?" He fully expected to get an answer, and he did.

She said, "I think the purpose of life is to make the planet a better place when you leave than it was when you got here. And to be kind and loving to all creatures—and that includes the plants, and the animals, and the rocks." Lastly, she said, "And I think the purpose of life is to remember love." She didn't say "find" love, but "remember" it.

This made us think about how much we all may know before we pull down the veils to protect ourselves from the outside world. We asked, "Jessica, where did you learn all these things?" And she said, "I just knew them before I came here."

Jampolsky: We believe that Jessica's story suggests that we all know about our purpose, but it becomes buried from our awareness. Diane and I have found from our audiences that one way to examine this idea is to ask, "If at the end of your life there is going to be one sentence to describe your life and how you spent it, what would you want it to be?" Ponder that, and then ask yourself if what you said is in harmony with how you're living your life. If not, you may want to take a look at your priorities and shift them around.

Perhaps a personal vignette might be appropriate. When I was about 14 years old, a friend of mine was killed in an automobile accident, and I lost any kind of faith I had in God. I became a militant atheist.

In high school, I remember writing a paper about wanting to be a physician. Part of the reason I wanted to be a physician was because of Dr. Schweitzer. I identified with what he was doing, but I was an atheist and his spirituality, to me, was an abstraction. I couldn't identify with it.

Then, in the late 1960's and early '70's, I became very depressed. I really didn't know why until, later, I found out from Mother Teresa that it was one of the biggest diseases of the world, she thought: spiritual deprivation. That feeling of emptiness inside, of not being connected to my source or to others. I had become an alcoholic. My inner world was in shambles, although to the outer world I was a successful psychiatrist.

It was shortly thereafter that I was given a book called *A Course in Miracles* about spiritual transformation. After reading just one page, I had a most unusual experience; I heard a little inner voice saying, "Physician, heal thyself; this is your way home." I knew that my life was going to be different, that it was going to be a life of giving.

It was then that we began to conceive a Center for Attitudinal Healing for children who were dying of cancer. We began having support groups for children, their siblings and their parents. We began seeing adults and children with AIDS, and another group of people who just wanted to apply attitudinal healing principles to help change their lives. As a volunteer at the Center, I began to get a little inkling of what Dr. Schweitzer's life was about, what his spirituality was about.

Cirincione: For many people, Dr. Schweitzer's life remains an abstraction. One of the fortunate aspects of our lives has been the opportunity to meet many people, living today, who exemplify the things that Schweitzer stood for, and who have been inspired by his life. We see him through them. These people are not abstractions.

One such person is a young man who came a few times to one of our groups for adults with AIDS. He had lost his job because he had been sick so much. He had little money and was feeling despondent. But one day, standing in line at Burger King, he found someone worse off.

In front of him was a woman who lived in the streets, who lived in her clothes— the only clothes she owned. He remembers thinking, "What is she doing here?" And then, "Why, she's doing the same thing you are; she's getting something to eat." Then he heard the man behind the counter asking for 77 cents more, please. Obviously, the woman was short of money; she just didn't have it. But the young man heard a little voice that said, "Give her the 77 cents and you will always have all that you need."

He found himself digging in his pocket, pulling out 77 cents, and putting it on the counter. The man took the money; the woman took the bag. She didn't turn around, didn't say anything, just walked out of the restaurant. Then the young man went up to the counter and placed his order. And as he was standing at the counter, waiting for his burger and fries, his life changed. He knew now that he was going to live a life of giving, a life truly dedicated to helping others.

As he left the restaurant, feeling incredibly elated, the woman was standing there, looking straight at him. She didn't say a word, but they were face to face. And he says he knew that if there was a presence of Christ here on earth, it was inside this woman. He believes she came there to open his heart, and the experience did completely turn his life around. He began to volunteer his time. Instead of worrying about his own time, he created a home for others who were in a worse situation—homeless, or sick with AIDS and also homeless.

He teaches us that at any moment in life, there is an opportunity to learn. This story reminds me of one of our other speakers, Dennis Weaver, who has worked with Love Is Feeding Everyone (LIFE) in southern California, and who has inspired all of us to make sure that we never turn a corner without noticing who is sitting, sleeping or standing on that corner. I think Albert Schweitzer did the same.

Jampolsky: Another one of our teachers, a man named Henri Landwirth of Orlando, Florida, received the National Caring Award in 1988 in Washington, D.C.

He was a prisoner of war in Auschwitz at age 12. His parents were killed in the war, and many times he was nearly killed. After World War II, he came to the United States with only twenty-five dollars in his pocket. He ended up going to hotel management school, went to Florida, eventually owned several hotels and became a very wealthy person. But he still was not a very *happy* person.

He remembers one day asking for help from the universe and he got the message that what he needed to do was use his money to help children who were dying of cancer. So he started to bring sick children and their families to his hotels. He got Disney World to open up their facilities for free. He did this for about three years, and then there were so many children coming that he decided to build a village called "Give Kids the World." Today, there are 2,200 families coming to the village annually. And Henri is a transformed man. He wasn't able to find peace until he was able to live life in the present—not going over the past, not dealing with the future, but really living in the present and doing his best to be helpful to humankind.

Cirincione: We have the joy of being international ambassadors to this children's village, bridging other countries with the village so that their children can also come with their families. Henri has given up his personal business enterprises, turning them over to others, so he can work full time in the village.

Another dear friend who is carrying the Schweitzer beacon of light is a woman named Zalinda Carusi. She wrote us a letter from Boise, Idaho, about two years ago asking for our help in starting a Center for Attitudinal Healing at the Boise State Penitentiary.

Zalinda went on to say that her 19-year-old son, John, was shot to death by a casual acquaintance who had been under the influence of drugs. The acquaintance, Michael, was sent to prison but periodically came up for parole. At first, Zalinda and her family were doing everything possible to keep him in prison for the rest of his life.

Jampolsky: But Zalinda was having a lot of problems. Her hair was falling out; she had serious skin disease and gallbladder problems. She felt that her hate, her anger and her feelings for revenge were killing her. This reminds us what attitudinal healing is all about. It's really not the people or conditions in our external world that cause us to be upset, but our own attitudes. And we can do something about them. We need to find inner peace for ourselves before we can bring peace to the world.

After getting her letter, I flew to Idaho and met with Zalinda and with the prisoner, Michael. It was the most amazing experience of my life to see her with absolutely no animosity. She had read one of my books, *Love is Letting Go of Fear*. She discovered that one needs another way of looking at the world rather than holding on to anger. Although she surely didn't feel like forgiving Michael, she wanted to listen to her inner voice, and that voice said to start visiting him, which she did.

She told him, in all her anger, just how she felt. But she kept going every week, and eventuallly her perception of Michael began to change. She met his parents and saw what agony they went through. Things changed to such a degree that Zalinda began to go to the parole department to get Michael out of prison, and on July 17, 1989 he was released. Zalinda was there to take him home.

Cirincione: The last teacher we would like to tell you about is Jennifer, a young woman we met when she was 11. She had failing kidneys and was on dialysis most of the time. But whenever you'd visit her in the hospital, you'd find her consoling a nurse who had lost a young patient, or helping someone who had just come into the hospital and was terrified. Her life was a life of giving. She found joy in bringing an incredible amount of love to every single thing she did.

When she was about 17, she was on a television show, and she said, "If I knew I had only one week left to live, I would think of all the people I had a grievance against, or people I hadn't forgiven, or those who hadn't forgiven me. I would write them, or call them, or at least think of them. I think before we leave here we are supposed to clean up all our unfinished business. And that means, heal our relationships."

Jennifer knew on a very personal level that the wars we have with other countries are the wars we have with ourselves and with our neighbors. How can we hope to get along with other nations if we can't get along with our own families?

When she was 20 years old, Jennifer died. At her funeral, her mother carried a card with something that Jennifer said just before she died, "I would rather have lived my life as a sick person helping others, than as a healthy person, only for myself."

Jampolsky: We would like to end our remarks with a short prayer from *A Course in Miracles*. We find it is a very powerful prayer that does wonders in getting our egos out of the way, allowing us to have a consciousness of giving and loving. In another way, we feel this prayer demonstrates what Dr. Schweitzer's life was all about:

I am here only to be truly helpful.

Cirincione: I am here to represent You who sent me.

Jampolsky: I do not have to worry about what to say or what to do.

Cirincione: Because You who sent me will direct me.

Jampolsky: I am content to be wherever You wish, knowing You go there with me.

Cirincione: I will be healed as I let You teach me to heal.

Albert Schweitzer:
Current Lessons for International Health

By William H. Foege

Forty years ago, as a teenager, I was confined for months in a body cast, but my mind was freed by the discovery of Albert Schweitzer. Thus I was allowed, during these months, to share his emotions: the shame when he killed a bird, the moment of his discovery of how richly blessed he was and what that required of his life, his understanding that the benefits of science and medicine must be shared with those who do not have them. So over the years I read what he wrote, and wondered about Africa. In August of 1965, 25 years ago this month, I went to work in a medical center in Africa. I somehow expected that Schweitzer would escape the bonds of mortality and that I would meet him. It was not to be . . . he died within days of my arrival in Africa.

On the other hand, it is precisely because of his immortality that we gather here today. As we've heard, his direct influence on the lives of people who lived within traveling distance of his hospital was great; even greater is the influence he has had upon people who did not physically stand in that hospital.

He never stopped growing through his life, but the exciting aftermath is what we've just heard: that he continues to grow in death, through the work of people who he inspired by his example.

We each enter the river of life at a given point, with the accumulation of history upstream, and we build on the experiences of the past. We take short cuts as we benefit from what went before. And we learn in a week, and sometimes in an hour, a theory or an explanation that might have taken a great person of the past an entire lifetime to untangle. We thrill to the increase in our functional life expectancy as we not only leapfrog to new applications, but develop greater efficiency because of science and technology. And in the midst of that I am struck by the productivity of Schweitzer, who was forced to use ships instead of airplanes, letters instead of fax machines, pens instead of word processors.

In 1977, Jonas Salk gave a talk in New Delhi entitled, *How to be a Good Ancestor*. We are here because Schweitzer was a good ancestor. Many will have their names in history books 115 years after they were born, but how many will have such a gathering of celebration as we are seeing today? And what does that 115 year legacy have to tell us about the challenges that face the world today to improve health? His practical medical lessons continue to be so relevant and so timeless that I have to remind myself all the time that he lived in a different age, an age when the sun did not set on the British Empire, while today the sun doesn't set on Hewlett Packard or the World Health Organization. In the year of Schweitzer's birth, Alexander Graham Bell was doing his pioneer work on the telephone, transportation for most people was no faster than it had been at the time of Christ, and health was a gift that could be snatched at any moment by microenemies that people didn't even imagine. For instance, in the year that Albert Schweitzer was born, in Fiji, with 150,000 people, they lost more than a fourth of their population—40,000 people—because of a measles epidemic. The state of our environmental concern and our reverence for life can be illustrated by a veto that year by President Grant of a bill calling for the protection of the buffalo from extinction.

Allow me to mention three aspects of Schweitzer's medical legacy with direct relevance to international health today.

The first, his combining of Science and the Humanities. We hear repeatedly about the art and science of medicine. But what does that really mean? Huxley once defined science as common sense at its best. And well-practiced medicine is indeed that. But it is possible to practice common sense medicine without ever touching the spirit, or exercising the creativity of art. Schweitzer, for me, demonstrates the composite of art, science and spirituality, with his creative, humane, common-sense-at-its-best medicine.

Will Durant has pointed out that the first person known in history as an individual, by name rather than by title, was a man by the name of Imhotep. He was an artist and a scientist. He was a physician. He founded a school of architecture in Egypt. He designed the oldest Egyptian structure extant—the Step-Pyramid. And over 5,000 years later, the combination of artist and scientist is still to be sought but rarely achieved. But as in a relay race, Schweitzer took that baton to demonstrate to a modern age how that could be done. And we heard in the previous talk how his music provided an example of harmony, which in turn became the metaphor for how he integrated art, theology, philosophy and science into a single harmonious chord: his concept of reverence of all life.

A second lesson for medicine of this and future generations: he taught us the importance of what we now call the global village; the linking of people medically and theologically in both time and space. This is an old concept; it was well-developed over 2,000 years ago by an historian, Polybius, who wrote, "It may have been possible in the past to have unrelated events, but from this time onward history becomes an organic whole." Everything effects everything. Polybius went on to say, "The affairs of Italy and Africa are connected with those of Asia and Greece." And of course we are reminded of that in this time, as we look at the Persian Gulf, and we wonder about the aftermath. And we again read the words that are in the program from Schweitzer: that we don't possess the superhuman reason which should accompany our superhuman might.

This unity of people in time and space requires a new way of thinking. For instance, we are dependent on those who have gone before, which Schweitzer appreciated in all of his studies. But we have a responsibility to push that body of knowledge, to add new thoughts and experiences, to leave the reservoir—this endowment of knowledge and understanding—better than we found it. And this he did in music, in the interpretation of Bach, the building of organs, the understanding of the historical Jesus, the practical application of tropical medicine . . . he expanded the capital for others to draw on.

But he taught us not only to build the endowment but to share it. Schweitzer's understanding of the relationships shared by those who went before, but also the lateral relationships with everyone around the world, helped propel the modern understanding of international health. His example has now been so institutionalized that we forget what a pioneering effort it was. We live at a time when the World Health Organization binds us in global objectives. We live at a time when AIDS forces us to recognize that what happens in Kinshasa makes a difference in New York. We live at a time when the United States has two importations of measles every week from other countries. The interrelationships seem obvious. We live in a time when the world has collaborated, both across the board and across the globe, to get rid of smallpox. We live at a time when the world expects to get rid of polio within this decade. We live at a time when the collaboration has been so good that this year 2.5 million children will not die because of vaccine-related diseases. Over a million children will not die because of diarrhea. And we have seen these, not as three-and-a-half million statistics but as three-and-a-half million faces: children who are allowed to smile across the dinner table tonight, children who can sit on the laps of seven million parents. We live on the eve of the largest gathering of heads of state to ever attend a meeting on any subject; the Summit for Children will be held in this building next month. And it is expected that over 70 heads of state, including President Bush and President Gorbachev will be here, to ask about the future of children.

So we may not appreciate in that environment the pioneering feat of a European sharing his knowledge with a small rural area of Africa . . . for their benefit. Missionary societies had of course sponsored physicians before. But primarily to minister to other missionaries or to use medicine as a means to proselytize. Schweitzer brought medicine to Africa to practice his beliefs, as a logical expression of his beliefs, not to force others into his beliefs. We are reminded of the words of William Penn, "To help mend the world is true religion." It is harder to overstate the importance of forcing medicine to overcome this barrier of geography. Because you see, medicine and public health have two barriers to good decisions, time and distance. The further the distance between a decision and its effect, the harder it is to make a good decision. But Schweitzer demonstrated that distance could be overcome, and one of his contributions to health today is to remind us that social justice requires that all must benefit from our knowledge, regardless of where they live.

But the global village also requires a concern for the future. Again, we may not appreciate the world that Schweitzer lived in because we now know about the ozone layer, the destruction of the rain forests, the population pressures in much of the world, the danger of weapons and the cost of weapons. This year the world will lose an estimated 15 million children before their fifth birthdays. UNICEF and others have estimated that over half of these deaths could be prevented at a cost of three billion dollars per year. Not many could foresee a world that would argue about whether it can invest three billion dollars a year into child care while spending that amount each day on weapons! Sam Levinson once said that it is not hard to be wise: "Think of something very stupid and do the opposite." Well, it is very stupid to spend three billion dollars per day on weapons.

And as we've heard today, Schweitzer saw that future, speaking passionately about the risky world of weapons and chemicals. So Reverence for Life was a guiding principle for him. It becomes a guiding principle for the future, for endangered species as well as endangered people. Reverence for Life promotes the idea of being a global citizen first and a nationalist only second. It deals directly with that second great barrier to good health decisions, time. The longer the time between a decision and the effects of that decision, the harder it is to make a good decision. It is easy to make the right decision if the effect is immediate, as in the use of antibiotics for pneumonia or a cast for a fracture. But it is difficult to make good decisions when the incubation period is long. That is why it has been so hard to quit smoking, because the effects won't be seen for decades. But it is even harder to make good decisions when the incubation period extends beyond our lifetime. For example, it is hard to make good decisions about nuclear wastes that may cause problems in hundreds of years. But Schweitzer reminds us, urgently, about reverence for future life.

In a year of great health advances, those 15 million children who I said will die before their fifth birthdays might just as well have been born 300 years ago, or 500 years ago, or a thousand years ago because they will not benefit from our 20th century medicine, they won't benefit from our science, they won't benefit from our computers or our airplanes. So the health challenge is for the heads of state, meeting here next month, to provide a world that is safe for present and future children.

• Their challenge is to agree on some expressions of security that are meaningful for children. No child, for example, need go through life crippled from polio, when prevention costs pennies. A polio vaccine costs two cents on the international market. Children need no longer die from measles; a measles vaccine costs six cents on the international market. Nor do mothers need to die by the hundreds of thousands a year in pregnancy. We now know how to achieve these ends.

• Their challenge is to agree on a transfer of resources, efforts and attention, from weapons to health and education.

• Their challenge is to agree on the use of our tremendous research capacity, not for more efficient and lethal weapons but for more efficient and effective vaccines.

• Their challenge is to agree to support the United Nations agencies, the World Health Organization and UNICEF, and not withhold funds, when they get upset, as a sign of displeasure as our country has done. It is an embarrassment that the United States has not paid its toll to the World Health Organization. Last year I wrote an editorial on this. I quoted Dolly Parton who said, "You would be surprised how much it costs to look this cheap."

• Their challenge is to elevate Reverence for Life, social justice and global citizenship to policy determinants rather than academic discussion.

The third and last contribution I will mention, as Schweitzer continues to teach medicine today, is his example of sharing the burden. He found health solutions in sharing problems rather than preaching from a pulpit. In his words, "It struck me as unconscionable that I should be allowed to lead such a happy life while I saw so many people around me wrestling with sorrow and suffering." Or as Gandhi said, "The golden rule is really to refuse to have what millions cannot."

During the smallpox eradication program, a man argued passionately with me that smallpox is nature's way of controlling population; we should not try to stop it. I asked to see his arm; he had a vaccination scar. I told him he had lost his moral authority. Because it is easy to make judgments for others if we don't have to share their risks.

We can't hoard our knowledge for our benefit alone. As Primo Levi said, "Once we know how to reduce torment and don't . . . then we become the tormentors." So we must share the burden. And just as with the first rule of medicine, which is "do no harm," we must strive to avoid increasing that burden. Schweitzer talked about people being diverted by the passing treasure of this world. But could he have imagined people systematically increasing the disease burden of the world in a trade for money? It is an irony of history that the R.J. Reynolds Company was born the same year as Albert Schweitzer, and that 1913 saw the introduction of both the Schweitzer Hospital and Camel cigarettes. The deliberate increase of the burden of suffering mocks Reverence for Life.

Schweitzer's lesson of "sharing the burden" leads us finally to ask, "Is there a measure of how well civilization is doing?" There is. The final criterion for measuring civilization is not found in knowledge, or science, or technology, or even happiness, or legal solutions, or freedom or health. The final measure of civilization turns out to be how people treat each other. It becomes a measure of a country, a company, an organization, a university, a person.

If we are to face the challenge of closing the gap between the health status of Africa and of the West, the scientific, logistic, resource and administrative demands will be daunting. But the hardest challenge will not be those. The formidable challenge will be: will we meet that Schweitzer test of being civilized people and civilized nations by how we treat others?

We pay homage today to Albert Schweitzer for living the definition of a civilized person.

Albert Schweitzer, Physician

By John A.M. Karefa-Smart

Unlike several of my fellow participants at this international symposium on Albert Schweitzer, I did not have the experience of meeting and conversing with the great physician. During my final year in medical school at McGill University in Canada, I was so impressed by what I had read and heard about him that I thought it would be the best preparation, before returning to my own country to start my medical career, if I spent a year with him as an intern. I therefore wrote to Dr. Schweitzer asking for permission to work with him. I did not receive a reply to my letter. Since it was written from Canada during the active Atlantic Ocean phase of World War II, it is very probable that my letter did not reach Dr. Schweitzer, or that it suffered the fate of being among the trunkloads of letters that he had left unanswered, for one reason or another, when he died in 1965.

For many years I unquestioningly accepted the reason given to me by a former missionary in the Belgian Congo that I, an African, arriving at Lambarene as a fully qualified physician and a graduate of a renowned medical school, would have presented a situation that Dr. Schweitzer was not, at that time, prepared to accept. For this reason, although Gabon was in the region of the World Health Organization for which I was the regional officer for three years, I purposely refrained from including a visit to Lambarene in the many official visits that I made to the countries in my area from my office in Brazzaville. My only qualification, therefore, for accepting this invitation to participate in this symposium is that I did visit Lambarene after Dr. Schweitzer died, and I have been, for more than 15 years, a member of the Board of Directors of the Albert Schweitzer Fellowship. This, I trust, should dispel any suspicion that I do not greatly revere the man in whose honor we are assembled.

The opinions that I am going to express have been formed only by what I have read in Dr. Schweitzer's own writings, and by what those who knew him have written and spoken about him.

Dr. Schweitzer recounts that as a young man he decided that he would devote himself, until he reached age 30, to the study of theology, philosophy and music; then he would spend the rest of his life in direct service to his fellow men. It was therefore only in 1906, when he felt that he had achieved his first goals, that he began his medical studies, completing them by taking the state examination in 1911. Medical studies in most European countries in Dr. Schweitzer's time consisted almost entirely of lectures and some practical work in anatomy, chemistry and physiology laboratories. Clinical work with patients was practically unknown during a student's years.

The thesis that Dr. Schweitzer presented for his medical degree was, not surprisingly in light of his previous academic interests, *The Psychiatric Study of Jesus*. Before he went to Africa he took a short course in tropical diseases and, 10 years after, on a visit to Europe in 1923 to raise funds for his hospital, he took another course in obstetrics and one in dentistry to prepare himself for his work.

The decision to go to Africa to pursue the goal that he had set for the second phase of his life was influenced, Dr. Schweitzer recollected, by the figure of an African in a statue of Admiral Bruat in the town of Colmar. "The expression of thoughtful sadness," Dr. Schweitzer wrote later about the face of the African in the statue, "spoke of the misery of the Dark Continent." Arrangements were made with the Paris Evangelical Missionary Society, which gave Schweitzer permission to build a hospital on their mission site at Lambarene. He was also offered accommodations in one of their houses. On Good Friday, 1913, accompanied by his wife, a trained nurse, Dr. Schweitzer sailed for Africa.

When the Albert Schweitzer Bresslau (after Mrs. Schweitzer's maiden name) Hospital began in a vacant chicken coop, Dr. Schweitzer had made the beginnings of his missionary medical work in Africa. The chicken coop soon became inadequate, so Dr. and Mrs. Schweitzer moved a little further up on the right bank of the Ogowe River, gradually adding building after building under his direct supervision. The new hospital was, in fact, an African village, differing from other Gabonese villages only in the fact that all the houses were built of wood, instead of mud and wattle, and were roofed with corrugated iron sheets. When he won the Nobel Peace Prize, Dr. Schweitzer used most of the prize money to build a satellite village for his Hansen's disease (leprosy) patients and their families. This he named the "village of light."

Patients came to be treated from near and far and, later, from all parts of Gabon. A special pavilion was set aside for European patients from the trading community. The diseases treated by Dr. Schweitzer were those commonly seen in African villages in the tropical forest zone: sleeping sickness, malaria, tropical sores, infected wounds, parasitic worms, diarrhea and dysentery, heart trouble, and a large variety of skin diseases, including leprosy. Patients also came to the hospital with illnesses that required surgical treatment: mostly fractures, wounds of all kinds, hernias and scrotal elephantiasis. Women in difficult and prolonged labor frequently required obstetric intervention. All of these ailments Dr. Schweitzer treated by himself.

Dr. Schweitzer's overriding concern was to relieve pain and suffering, so he took with him on his first trip to Africa large supplies of all the medicines that were available at the time for the treatment of the diseases already enumerated, and for the relief of pain. These supplies were frequently replenished with funds generously provided by friends and admirers in Europe and in the United States. Adequate stocks of bandages and other supplies for surgical work were always kept. The gentle care that Dr. Schweitzer trained his assistants to give—which must have been a striking contrast to the impersonal care given at the government hospital on nearby Lambarene Island—was preferred by the patients, and word of this spread over Gabon and beyond.

Another attraction of the Albert Schweitzer Bresslau Hospital was the familiar village atmosphere. The families of the patients were allowed to cook meals on open wood fireplaces in any available space between the hospital buildings. Goats, sheep, cats, dogs and poultry were permitted to roam at will all over the hospital grounds, just as they did in the surrounding villages. At night, members of a family slept on mats on the floor beside each bed. Several visitors from abroad have recorded that, perhaps in deference to Dr. Schweitzer's philosophy of Reverence for Life, no active measures were taken to get rid of insects such as cockroaches, flies and mosquitoes. Bed nets, when available, were the only protection at night from the bites of malaria-carrying mosquitoes and from the tse-tse flies that transmit sleeping sickness. Gabonese patients therefore felt very much at home in Dr. Schweitzer's hospital.

Dr. Schweitzer wrote about "the joy of being here, working and helping." He also wrote, "This does not mean that I can save life—we must all die—but that I can save men from days of torture. Pain is more terrible a lord of mankind than death itself." A visiting American wrote after being at Lambarene, "One doctor with the most modest equipment can, in a single year, free from the power of suffering and death hundreds of men who must otherwise have succumbed to their fate."

From the early beginnings when he had only the skilled assistance of Mrs. Schweitzer and one or two Africans whom he trained as practical nurses, the hospital staff at Lambarene gradually increased in number. A nurse from Europe arrived with another doctor in 1924. A third doctor and another nurse joined the staff in 1925. In subsequent years, several doctors and nurses came to Lambarene, some for short, and others for longer, periods. They came from Germany, Switzerland, England, the United States and several other countries where Albert Schweitzer Bresslau Hospital support groups had been organized.

As other physicians became available, Dr. Schweitzer gradually relinquished treatment of his patients to his assistants and devoted more and more of the time left after his writing and his music to the planning and supervision of construction of additional buildings, and work in the hospital village. He was, it is reported, especially hard on any Africans who did not share his enthusiasm for unpaid manual labor. He told the story himself of how he once lost his temper and loudly exclaimed, "What a fool I am that I have tried to be a doctor to such savages!" To which his trusted Gabonese assistant Joseph replied, "Yes, Doctor, on earth you are a big fool, but not in heaven."

Not all who went to Lambarene to work with Dr. Schweitzer have written flatteringly about him. One of them, an American doctor, wrote in 1963, "During the last 20 years the doctor has practiced very little medicine or surgery himself." Another wrote, "Those who have actually seen Dr. Schweitzer in the performance of his medical duties are rare birds indeed. The simple fact is that Schweitzer's medical work has been at a minimum." Another visitor wrote, "I saw him touch one patient during nearly two years in the hospital."

Dr. Schweitzer, however, cannot be blamed for appearing not to have much understanding of, or concern about, what we now know as public health. When he first went to Gabon there were no schools of public health either in Europe or in America. In Gabon, the control of "les grands endemies" was the responsibility of the French colonial paramilitary organization, S.G.H.M.P., which conducted mobile campaigns against epidemics, such as cerebrospinal meningitis, and administered such vaccinations as were available. Little opportunity was left for Dr. Schweitzer himself to engage in this preventive public health activity.

Hospital physicians in Gabon, as in other African countries, provided only outpatient or inpatient care. The Albert Schweitzer Bresslau Hospital was not alone among missionary hospitals all over the world that saw their primary task as bringing relief from pain and suffering from a clinical or hospital base. It was only a few years before Dr. Schweitzer died that the Christian Medical Commission of the World Council of Churches, of which I had the privilege of being a founding commissioner and vice-chairman, began to challenge medical missionaries in many countries all over the world to look beyond the confines of hospital compounds to the surrounding communities as the true places where the provision of basic health services would result in measurable and lasting improvements in the health status of men, women and children.

It must also be said that, to his credit, Dr. Schweitzer was not only concerned with the medical care of Africans. He also had very strong views on what he regarded as the evil effects of western civilization on the health of Africans. He would have been untrue to the time in which he lived, and to his background and culture, if he had espoused different views than those he held about relations between Africans and Europeans. As he wrote, the African was his brother, but only a very junior brother.

Dr. Schweitzer was also one of the few who early recognized that alcohol is a health problem—a view only very recently adopted by public health officials the world over. He wrote, "The most important thing is that we cry 'halt' to the dying out of the primitive and semi-primitive peoples. Their existence is threatened by alcohol with which commerce supplies them, and by diseases that already existed among them, but which, like sleeping sickness, were first enabled to spread by intercourse which civilization brought with it." It can therefore truly be said, in tribute to him, that the World Health Organization, the U.S. Secretary of Health and Human Services, and the U.S. Surgeon General are all following Dr. Schweitzer's pioneering leadership in the important campaign against alcohol.

It is important that we now admit that prevailing modern concepts about effective medical practice and about public health are not suitable yardsticks with which to measure the achievements of someone like Dr. Schweitzer who set his own goal and faithfully followed it. At the beginning of his medical career, Dr. Schweitzer had set himself the goal of doing what he could to bring relief from pain and suffering to his patients in Gabon. No one can doubt that he did that. He made no claims to greatness as a physician or surgeon. He simply followed the example of Jesus of Nazareth, another compassionate person who went about doing good and healing the sick. The gift of compassion gave Dr. Schweitzer the power to heal.

Many critics boldly prophesied that the Albert Schweitzer Bresslau Hospital would not survive Albert Schweitzer's death. Today, the hospital not only survives but has entered into a new phase of life trying to enlarge the vision of its founder. The hospital has assumed a new mission of working not only to treat individual illness, but also to help the villages to gradually become healthy communities.

Other critics say that Albert Schweitzer is a myth. They have not taken the trouble to see the vision about which he wrote in his *Mitteilungen aus Lambarene*: "On the brotherhood of those who bear the mark of pain lies the duty of medical work for humanity's sake in the colonies. Medical men must accomplish among the suffering in far-off lands what is crying out for accomplishment in the name of true civilization."

We must all seriously weigh the words of Norman Cousins who wrote in 1960, "If Albert Schweitzer is a myth, the myth is more important than the reality. For mankind needs such an image in order to exist. People need to believe that man will sacrifice for man, that he is willing to walk the wide earth in the service of man. It would simplify matters if Albert Schweitzer were totally without blemish, if his sense of duty toward all men carried with it an equally high sense of forbearance. But we cannot insist on morally symmetrical ideas all the time. We can derive spiritual nourishment from the larger significance of his life as distinct from the fragmented reality."

Albert Schweitzer, the very human, stubborn and overbearing, but gentle and compassionate "grand docteur" of Lambarene has earned for himself a place among the immortals. The very least that we can do in his memory is to honor him as we are doing at this International Albert Schweitzer Colloquium.

Health Care in Exploited Societies

By Victor W. Sidel

Dr. Schweitzer's medical work, starting in 1913 in Lambarene in what was then French Equatorial Africa, saw the development of his clinic from a converted chicken coop in a clearing on the banks of the Ogowe River to a multi-building facility treating hundreds of patients with acute and chronic illnesses and sheltering large numbers of patients with leprosy.[1] On the lamp outside the Schweitzer Hospital was the inscription, "Here, at whatever hour you come, you will find light and help and human kindness."[2] Others in this symposium have spoken movingly of "his gentle and compassionate" nature, of his attempts to relieve the pain and suffering of his patients and their families, and of the "spiritual nourishment" that can be derived from "the larger significance of his life."[3]

I plan to discuss briefly three aspects of the relevance of Dr. Schweitzer's work and of his philosophy to problems in medicine and health care that we face at the dawn of the 21st century. These aspects are:

- The relevance of Reverence for Life to medical decisions on life and death, particularly in industrialized countries;
- The role of medicine in the empowerment of exploited and powerless people in both industrialized and less industrialized countries; and
- The importance of the redistribution of resources being used for arms to health and development, again in both industrialized and less industrialized countries.

With regard to the relevance of Reverence for Life in technologically-advanced countries, the qualities of humane medical care that Schweitzer expounded and tried to practice are often subordinated to efforts to supply patients with the advanced technical tools for their diagnosis and treatment. Indeed the life-prolonging potential of some of these tools is so great that learned philosophers, lawyers and clinicians literally spend hours debating the appropriate ethical initiation or continuation of such methods. Dr. Schweitzer's principles must, I believe, be repeatedly reinterpreted in situations in which prolongation of life represents prolongation of a lingering and painful death. When does Reverence for Life demand *removal* of life-supporting devices or of artificial feeding tubes? When does it demand that the physician assist the patient in suicide, as occurs openly in The Netherlands and much more surreptitiously, but not infrequently, in the United States? It is reported that Dr. Schweitzer himself, at the time of his death, refused death prolongation measures.[4]

Another aspect of Reverence for Life concerns us: May a society, especially one that evidences little concern for the quality of life of many of the babies being born into it, and offers little incentive or assistance to women and men in the prevention of pregnancy, limit the reproductive choices of its women? Do such limitations represent Reverence for Life or a debasement of the life choices of women? What is the role of physicians and other health workers in such choices?

More generally, the classic role of the physician, as Dr. Schweitzer knew it, centered on the dyad of "authority" and "caring"; indeed these principles represent the qualities for which he was so universally admired. But because of increasing concern with the complexity, power, cost, maldistribution and frequent inhumanity of much of current medical care, these principles of authority and caring are now often perceived as necessary but not sufficient to define the work of the physician. They must now be joined by a third principle, that of empowerment. If, as W.H.O. defines it, "health is a state of complete physical, mental and social well-being," it is hard to see how a health professional can strive for the health of patients without empowering them—with specific technical skills and tools when necessary—to make effective decisions about the continuation of their pregnancies, of their own treatment or even the continuation of their own lives.

This brings us to our second topic, the role of medicine in empowerment. The meaning of empowerment goes beyond the enhanced ability of the individual or family to determine their reproductive choices and to decide on the nature and duration of their own treatment and even of their own lives. For the promotion and preservation of health, empowerment must extend to the ability of people, singly and in community, to change as much as possible the social, economic, cultural and environmental conditions that, to a large extent, determine the quality of their lives and the level of their well-being.

If the individual is rich, there is usually much that he or she can do to change surrounding social, cultural, economic and environmental conditions, although the pervasiveness of culture and the largely unyielding nature of the environment may at times defeat even the most powerful individuals. If the individual is poor, there is usually little that he or she can do to change the conditions of life. Only by social mobilization can individuals gain enough power to change social and economic conditions and to attempt to gain significant influence over cultural and environmental conditions.

Individuals in poor societies usually suffer under a dual powerlessness, both their own and that of their societies. The existence of these groups, which include not only people in the Third World but also large numbers in rich countries such as ours, is largely based on past and continuing exploitation. Their labors or resources have created the wealth of the affluent, and continue to increase that wealth. In this context, the role of health professionals must include attempts to empower exploited communities to change their relationships to power structures that cause them to live in illiteracy, misery and sickness while the rich become even richer. Indeed, if our goal is improvement in the health of people in poor communities, Schweitzer's principle of human kindness, while indispensable, is not enough. The light and help he offered must include teaching people how to join together to change the conditions that produce illiteracy, misery and sickness.

It is reported that one of the influences that helped Schweitzer in his decision to become a physician in Africa was a 1904 article calling for the services of qualified people to assist in the work of the Paris Mission Society in what was then the Congo.[5] Missionary physicians, like physicians sent by the companies or the governments of imperialist societies, usually had dual agendas in the areas to which they were sent. One was indeed to help treat the sick people of the region; but the other was to participate in the colonization of the region, either for the greater glory of God or for the commercial or national interests of their sponsors. Schweitzer, with his fundamental suspicions of both organized religion and imperialism, presumably did not relish being viewed as their agent; but that did not save him from obeying the wishes of the colonialist government that permitted him to maintain his clinic nor did it keep many people of the region from harboring deep suspicions of his motives and methods.[6]

The empowerment of people must of course include giving them opportunity and help in learning. That part of the role of the physician is embodied in the title "doctor," derived from the Latin word for "teacher." Even education limited to methods for survival under oppression can be useful. For example, there are excellent data showing the close relationship between breast feeding and child survival and between female literacy and decreases in infant mortality rates. While the teaching of skills for coping more successfully with life under hard conditions is extremely important, education that leads to understanding of the nature of exploitation and oppression, and of methods for attempting to lift them, are of even greater importance in the struggle for health.

Examples of attempts to do both simultaneously—what has been called "revolutionary pedagogy"—may be found in Paulo Freire's efforts to develop methods to help people to perceive their personal and social reality more clearly and to deal critically with it. Transformation to a radical self-awareness may help people to reject passive responses to conditions of life and to take responsibility, with others, to change the oppressive and exploitive structures of society.[7] Mao Zedong and his associates attempted to do that in China from the 1930's to the 1960's and, although there were some initial successes, had their policies for empowerment later criticized and largely reversed. Under the government of Dr. Salvador Allende in Chile in the early 1970's, many health workers attempted to work for the empowerment of "*el pueblo unido*" and were rewarded—after the military coup that overthrew and killed

President Allende—with imprisonment, torture and even death.

The physician's role in education—even radical education—is not enough. The role of the physician must include advocacy of specific measures for change. In a narrow sense the advocacy should press for improved methods of public health, such as the methods advocated by UNICEF to reduce childhood morbidity and mortality. In order to be truly effective, however, advocacy must press for structural change. It must call for an end to an international economic order that forces poor societies to sell their resources or their labor cheaply and to buy manufactured products dearly. It must call for a reversal in the diversion of resources to arms, in both industrialized and poor countries, from development and human services. It must call for protection of the global environment, without differentially handicapping the industrial development of poor countries and thus permitting the rich to continue their advantage over the poor. Without in any way disparaging the contributions that Dr. Schweitzer made to the lives of his patients and their families, one must wonder if his impact might not have been even greater had he used more of his estimable talents for radical education and advocacy as well as for humane medical care.

However that may be, we as physicians are today faced with our own choices. Most of us lack Dr. Schweitzer's knowledge, skill and worldwide reputation and are therefore largely powerless as individuals, but by mobilizing groups of like-minded individuals our individual efforts can be magnified. Two such groups, about which Dr. Leaning spoke this morning, are Physicians for Social Responsibility (PSR), which Dr. Bernard Lown and a group of Boston physicians founded in 1961 and which Dr. Helen Caldicott was instrumental in revitalizing in the 1980's. In 1980 Dr. Lown and the U.S.S.R.'s Dr. Evgeni Chazov, both leaders in world cardiology, founded the International Physicians for the Prevention of Nuclear War (IPPNW), a federation of physician groups similar to PSR.

In the quarter century since the founding of PSR and in the decade since the founding of the IPPNW, those of us involved have been studying the medical and social consequences of the use of nuclear weapons. The lesson we have learned and have widely taught is that, since doctors can do almost nothing to deal with the physical, medical and social destruction that would result if the weapons were used, doctors must—by analogy with other medical problems that cannot be successfully treated once once they occur—work actively for the *prevention* of nuclear war. Indeed it was for this work, largely on topics covered this morning, that the Nobel Peace Prize was awarded to IPPNW in 1985. Physician groups affiliated with IPPNW have now been organized in 77 countries and Dr. Leaning this morning has told us of some of the current work of PSR and IPPNW.

While prevention of nuclear war remains our central goal because of the irreparable devastation it would cause, over the past few years many members of PSR and of the other national physicians organizations affiliated with IPPNW have been writing and speaking on a closely-related topic: the impact of the arms race on the health and well-being of the people of the world even if the arms themselves are never directly used to kill or maim. These consequences, which have become known as "destruction before detonation,"[8] tie the topic of this session to this morning's topic and also lead us to the third issue I want to discuss. One aspect of the destruction is the radiation released—and the consequent injuries—caused by nuclear testing and by purposeful and accidental leaks of radiation at nuclear weapons production plants. PSR and many other groups in the early 1960's—with the active participation of Dr. Schweitzer—worked for the adoption of theLimited Nuclear Test-Ban Treaty. The U.S. and the U.S.S.R. have stopped atmospheric testing (although France and China still continue such tests) but underground testing, with evidence of continued venting of radionuclides into the atmosphere and concentrations underground, has continued in the U.S. and the U.S.S.R. Furthermore, the releases of radionuclides at nuclear weapons production plants have been hidden by the U.S. government, by the Soviet government, and probably by other governments for years, and adequate studies of the health consequences in surrounding populations have never been done.[9]

Another aspect of "destruction before detonation," one directly relevant to Dr. Schweitzer's work in Africa, is the diversion of resources from development to arms and from human and health services to arms.[10] Between 1960 and 1985, an estimated $15 trillion has been spent on the world's military forces; today, military expenditures have climbed to over $900 billion each year, an amount in constant dollars about two-and-a-half times the level of 1960. Since 1945, some $4 trillion has been spent on nuclear weapons alone.[11]

As we heard this morning, "national security" is often invoked as the reason for increased military spending. But the world's stockpiles of nuclear weapons already contain the equivalent of some three tons of TNT for every person on earth, ready to be used at any moment by accident or design. The overwhelming destructiveness of these weapons and the pervasive nature of their effects make them militarily useless. Such weapons do not increase national security; they simply put the world at greater risk. By devoting excessive resources to the military at the expense of a vibrant economy, sound education system, poverty relief, health care, and meeting other pressing human needs, the nations of the world have neglected—and indeed undermined—their true security.

Several of the world's industrialized nations, the United States and the Soviet Union in particular, spend large amounts of their resources on arms. The expenditures lead to general economic problems that affect health and human services as well as specific diminution of governmental funding for services to promote health and human welfare.[12]

In the United States, for example, annual military spending has risen from 140 billion dollars to 300 billion dollars over the past decade. The total spending on the military over these 10 years was well over $2 trillion dollars, over $20,000 for each U.S. family. Of these expenditures, it is estimated that 20 to 25 percent were spent on nuclear arms and their delivery systems. Federal budget deficits have soared along with military spending. The Reagan Administration's addition of 1.6 trillion dollars to the U.S. national debt, was greater than that of the Truman, Eisenhower, Kennedy, Johnson, Nixon, Ford and Carter administrations combined.[13]

Along with their general economic effects, the large amounts of tax revenue spent on arms directly divert these monies from health and other human services. Yet, beneath this veneer of wealth lies the reality of America's poor and destitute. In recent years, the gulf between the rich and the poor has only widened. Some 33 million people live below an unrealistically low "poverty line." Of every four children under the age of five in the U.S., one lives in poverty; of every two black children under the age of five, one lives in poverty.[14] Many more millions have incomes just above the poverty level, but can barely survive. Millions of children and adults in the U.S. are homeless and millions are hungry. Inadequate prenatal care, nutrition, support and immunization bring shame upon our health care system. Some 35 million Americans lack medical care insurance and another 20 million have medical care insurance so limited they can be bankrupted by a major illness. Private hospitals and physicians across the country turn away patients who are unable to pay for health care.

An even greater impact has been felt in the Soviet Union. Although the U.S. and U.S.S.R. have in recent years spent roughly similar amounts of resources on arms, because of its lower G.N.P. the U.S.S.R.'s percentage of G.N.P. spent on arms is considerably greater. It is estimated that in the 1980's the U.S. spent about seven percent of its G.N.P. for military purposes, compared to 12 percent spent by the U.S.S.R.[8] These massive expenditures have had major economic and health and human services effects in the U.S.S.R.

In the world as a whole, the over $900 billion now spent annually on arms is equivalent to two-and-one-half billion dollars each day, over $100 million dollars each hour, almost two million dollars each minute, thirty thousand dollars each second. The total of almost one trillion dollars is equivalent to the incomes of 2.6 billion people in the 44 poorest nations, over one-half the world's population. This one-year expenditure on arms also approaches the entire debt that the poor nations of the world owe the rich nations, the debt whose interest payments are strangling many of the world's poor countries and which is said to threaten much of the world's banking system. Furthermore, this extraordinary waste of the world's resources has been increasing at a rapid rate. In 1960, world military expenditures totaled approximately 4.7 percent of world economic output; the expenditures now amount to over 6 percent of world output.

This unforgivable waste of human and material resources takes place in a world that cannot stand by and permit it to continue. The many problems plaguing the Third World—overwhelming poverty, political instability, crippling foreign debt and human rights abuses—are exacerbated by the diversion of enormous resources to the military. In developing countries, it has been estimated that close to one billion people are below the poverty line, 780 million people are undernourished, 850 million are illiterate, 1.5 billion have no access to medical facilities, an equally large number are unemployed, and one billion people are inadequately housed.[15]

Some 15 million children die preventable deaths every year, one every two seconds.[16]

Of course, along with the diversion of revenue to support needed research, arms spending also diverts highly trained people from working to improve health and the quality of life to support military functions. World expenditures on weapons research exceed the combined spending on developing new energy technologies, improving human health, raising agriculture productivity and controlling pollutants.[17]

The gulf between developed and developing countries continues to widen. From 1960 to 1983, the gain in real per capita income was 12 times larger on average in developed than developing countries. While the per capita increase for the richest fifth of the world's population was $4,800 during the period, the average gain was $70 for the poorest fifth. The rapidly rising military expenditures of developing countries have far exceeded the slight rise in the foreign economic aid they have received since 1960. The developed countries, on average, now spend 5.4 percent of their G.N.P. for military purposes and 0.3 percent for development assistance to poorer countries; in other words, the developed countries of the world spend about 20 times as much on military programs as on foreign economic aid.

Compounding this tragedy is the fact that even small conversions of the funds being spent on arms into spending on health could produce enormous benefits.[8]

• The cost of one hour's world spending on arms is equivalent to the entire cost of the successful 20-year effort to eradicate smallpox from the earth.

• The cost of three hours' world arms spending would pay for all of the World Health Organization's annual budget.

• The cost of one day's world arms spending annually would cut in half the 15 million preventable deaths of children.

• The cost of three weeks of world arms spending annually would pay for primary health care for every child in the poor countries of the world, including safe water supplies, and full immunization and basic food supplies.

What can we do that would place us in the company of Dr. Schweitzer in the context of today's problems and today's knowledge? First, physicians and other health workers in every nation must urge their governments to reallocate funds from arms to human services and development in their nation and to economic development aid to poor nations. To give us hope, there are countries that have accomplished such shifts. Costa Rica has consistently over the past three decades spent extremely little on arms. It has in consequence been able to use these resources for improvement in its health and social conditions, and the results have been dramatic. In the U.S., as a result of the work of PSR and other groups, past surveys indicated that one-half the population felt that "too much" was being spent on the military,[18] but those views may have shifted in response to United Nations and, especially, U.S. actions following Iraq's invasion of Kuwait. The U.S. Administration had proposed a small decrease in military expenditures for 1991, but it was a reduction far less than the one urged by PSR and other groups. In the U.S.S.R., President Gorbachev has repeatedly stated the Soviet Union's eagerness to reduce expenditures on both nuclear and non-nuclear arms[19] and has begun to implement that policy.

But ending the diversion of resources to the arms race and assigning them to development and to human and health services is not enough. The report of the 1987 International Conference on the Relationship Between Disarmament and Development stated:

"The world can either continue to pursue the arms race with characteristic vigor, or move consciously and with deliberate speed toward a more stable and balanced social and economic development within a more sustainable international economic and political order; it cannot do both.[15]

The health workers of rich nations must, therefore, as a second task, urge their governments to permit change in the international economic and political order to meet the just needs and expectations of the exploited people of the world. Part of this effort must be a reduction in the overwhelming burden that debt service places on the poor nations of the world. If the burden of debt and the obscene maldistribution of the world's wealth and resources are not addressed, we can expect a never-ending series of military crises like the one we now have, and expect continued gross differences in health status between the rich and the poor.

Finally, the health workers of all nations must act to redistribute wealth and power within their own societies. More narrowly, they must demand that additional resources be allocated to ameliorate the causes of ill-health: poverty, illiteracy, homelessness, hunger and despair. Even more narrowly, they must demand and be a part of expanded programs of community development, health education, prenatal care, immunization and other forms of public health and preventive medicine.

Actions such as these would honor the memory of Albert Schweitzer, whose life embodied an epitome of authority and caring and whose philosophy cries out for an end to the oppression, deprivation, illness and hopelessness that is preventable only by the empowerment of the exploited people of the earth.

[1] Goldwyn, Robert. "Surgery at the Albert Schweitzer Hospital, Lambarene, Gabon." *New England Journal of Medicine*, Vol. 264, 1961: 1031-33.

[2] Straus, Maurice B., ed. *Familiar Medical Quotations*. Boston: Little, Brown, 1968: 220.

[3] Karefa-Smart, John A.M. *Albert Schweitzer, Physician*. Paper prepared for the International Albert Schweitzer Colloquium, United Nations, NY, August 23, 1990.

[4] Goldwyn, Robert. Personal Communication.

[5] Nobel Foundation, *Nobel: The Man and His Prizes*. New York: American Elsevier, 1972.

[6] Berman, Edgar. *In Africa with Schweitzer*. New York: Harper and Row, 1989.

[7] Friere, Paulo. *Pedagogy of the Oppressed*. New York: Seabury Press, 1970.

[8] Sidel, Victor. "Destruction before detonation." *Lancet*, Vol. 2, 1985: 1287-89.

[9] International Physicians for the Prevention of Nuclear War. *Radioactive Heaven and Earth*. New York: Apex Press, 1991.

[10] Sidel, Victor. "The arms race as a threat to health." *Lancet*, Vol. 2, 1988: 442-444.

[11] Sivard, Ruth Leger. *World Military and Social Expenditures 1989*. Washington: World Priorities, 1989.

[12] Sidel, Victor. "Socioeconomic effects of the arms race." Preventive Medicine, Vol. 16, 1987: 342-353.

[13] "Two trillion dollars in seven years." *The Defense Monitor*, Vol. 16 (No.7), 1987.

[14] Sidel, Ruth. *Women and Children Last: The Plight of the Poor Woman in Affluent America*. New York: Viking, 1987.

[15] *Report of the International Conference on the Relationship Between Disarmament and Development* (A/CONF. 130/39). New York, United Nations, 1987.

[16] UNICEF. *The State of the World's Children 1990.* New York: Oxford University Press, 1990.

[17] Brown, L.R. *Redefining National Security: State of the World 1986.* New York: Norton, 1986.

[18] *Americans Talk Security.* Boston: Marttilia and Kiley, 1987.

[19] Gorbachev, Mikhail. Speech to the United Nations General Assembly, December, 1988.

August 24, 1990

Lewis

Keynote Address
Our Future is Our Children

By Stephen Lewis

I would like to speak this morning to the ethical considerations attendant specifically on the rights of children. And that I will do as passionately as I can. Let me begin by providing a context within which I'll pursue the subject matter, a context that is perhaps simple and predictable. On September 29th and 30th, there will gather, here at the United Nations, the remarkable Summit on Children, the largest-ever assemblage of heads of state and heads of government in world history. We've had summits east and west, we've had summits north and south, but we have never had a summit of all the regions of the world with such remarkable numbers, likely upwards of 70 or 80 heads of state and heads of government. It will be chaired by the Prime Minister of Canada, in conjunction with the President of Mali.

And what binds all of the participants in this quite astonishing meeting, without precedent, is profound concern for the state of the world's children—often tragic, often desolate, often indescribably grotesque. And what is hoped is that from the meeting will emerge a series of goals and targets achievable by the year 2000 that will diminish the death, the disease, the suffering, the inequality and the injustice for the children of the world, and particularly for the children of the developing world.

There is, however, another underlying theme to the conference, another organic continuum that suffuses everything. And that is the Convention on the Rights of the Child. You will know that the Convention was passed unanimously by the General Assembly of the United Nations in November of 1989. It thus became the latest in the numbers of international covenants that become matters of binding international law when the requisite number of countries sign and ratify them. In the case of the Convention on the Rights of the Child, I think the process of ratification was faster than any other previous international convention. I was Ambassador for Canada to the United Nations between 1984 and 1988. I can remember monitoring the ways in which countries signed and ratified conventions. I can remember the response to the Convention Against Torture, which showed an unusual rapidity, but nothing to approximate the way in which the Convention on the Rights of the Child has been embraced. And so quickly did the requisite number of countries—20, in this case—ratify, that as of September 2nd, the Convention on the Rights of the Child has the force of binding international law on those who are covered by its provisions. I urge you to read it. Get it from UNICEF or any of the other organizations concerned with the fate of children.

It speaks profoundly to the rights of children everywhere in the developing and developed world; the rights of children in war; the rights of children as refugees; the rights of children subject to cruel, inhumane and degrading punishment; the rights of children trapped in child labor or prostitution; and the rights of children subject to egregious and demented physical, psychological and sexual abuse. It speaks also to the rights of children to health, to education, to a voice before the law, and rights of children in foster-family situations. There is barely a right that is not touched upon in the convention that doesn't speak to the needs of children throughout the world.

It is a convention of persuasive lucidity; not long, but compelling, uncompromising and principled. And it is a document that makes the intellectual juices churn just to read its contents. It is a stunning, encyclopedic recitation of the rights of the child. And there's not a country on the face of the earth that doesn't transgress one or many more of the articles, not a country that isn't in breech even as they sign and ratify. The ethical considerations that are raised are, in part, the subject of this colloquium. But in terms of the fundamental rights of the child to a full and decent life, let me give you the agonizing litany that tells the tale.

Every day, 7,000 children under the age of five die from measles, pertussis, tetanus and diphtheria because they were not immunized by a course of vaccines costing $1.00. Ethical values are shredded when the right of a child to a rich and vibrant life is sacrificed for the sake of $1.00. Every day another 7,000 children die from the greatest single child killer: dehydration induced by diarrhea because parents didn't have, or didn't know how to apply, the simple remedy of oral rehydration therapy—the combination of salt, sugar and water—an application costing seven cents. Ethical values lie in tatters when the rights of the child to a full and vibrant life are sacrificed for the sake of seven cents. Every day another 7,000 children die from respiratory infections, primarily pneumonia, because of the lack of antibiotics valued at $1.00. Ethical values are left in desecrated remnants when the rights of the child to a rich and vibrant life are sacrificed for the sake of a dollar. Every day, a thousand children in the developing world go blind because of the lack of vitamin A, whose cost, were it administered, approximates 10 cents. The very word ethics, in relation to human rights, loses its meaning when juxtaposed with that kind of human nightmare.

Statistical recitations can go on forever. One writer put it best with the phrase "the obscene daily harvest of children." In elemental terms, we're losing 14 million children a year, under the age of five, in the developing world from illnesses and diseases that are largely preventable by low cost, high impact interventions. And it is beyond belief that it continues in the fashion that I have recited. It contravenes every essential section of the Convention on the Rights of the Child. And the world, for whatever reason, is fundamentally impervious.

Forgive these observations: Iraq annexes Kuwait, and the world is rightfully filled with denunciations of immorality and mobilizes in response. There are clashes in the West Bank and the Gaza, and they are front page news everywhere. There are despairing and wretched fratricidal conflicts in the townships of South Africa and, quite rightly, it is a matter of ubiquitous public concern. But what is called by UNICEF "the silent emergency"—not these loud emergencies, but the silent emergency—of fourteen million lives lost every year . . . that is somehow greeted with inertia, even indifference. I will never understand it. It so offends human sensibilities that one almost chokes on the words in conveying it. Even more so because the response we have made in applications of immunization and oral rehydration therapy in the 1980's have already saved the lives of three million children. Why does that achievement not trigger a crescendo of excitement to save tens of millions more in the 1990's?

Albert Schweitzer, in whose name this colloquium is pursued, made a magnificent contribution, one of many contributions to the continent of Africa. And I want to dwell therefore, for a moment, on that continent in the context of the relationship of ethical considerations to human rights. I want to do it carefully, but in an unqualified fashion.

Subject to the predictable caveats and reservations, we know on the basis of thoughtful contemporary analysis that in order for the 45 countries of sub-Saharan Africa to reach the goals and targets for children in the year 2000 in health, education, nutrition, water and sanitation, would cost roughly seven-and-a-half to eight billion dollars *additional* per year.

Perhaps I can give you a sense of what the goals and targets are, so that you can understand the value of that investment to the human condition: a one-third reduction in under-five death rates. A halving of maternal mortality rates. A halving of severe and moderate malnutrition. Safe water and sanitation for all families. Basic education for children. A halving of adult illiteracy rates, and equal educational opportunity for males and females. The eradication of polio. The elimination of neonatal tetanus by 1995. A 90% reduction in measles cases and deaths. Achievement and maintenance of at least 85% immunization coverage of one-year old children, and universal tetanus immunization for women in the child-bearing years. A halving of child deaths caused by diarrhea, and a 25% reduction in the incidence of disease. A one-third reduction in child deaths caused by acute respiratory infections. There are further objectives relating specifically to protection for girls and women in matters of nutrition, in matters of education. What we are striving to achieve by the year 2000, in the context of the rights of the child, has never been within our embrace before, but it is now. And that is particularly true for the most vulnerable continent on the face of the earth, the continent of Africa. And within that continent, the 45 countries of sub-Saharan Africa. Again, by all the best estimates, were we to find the seven-and-a-half to eight billion dollars additional per year, and apply it to all those purposes—health, education, water, sanitation, nutrition—by the end of the year 2000 we would have saved the lives of 18 million children under the age of five in sub-Saharan Africa.

I want to point something out. In 1990 the sub-Saharan countries collectively paid nine billion dollars in debt-servicing obligations. In other words, we take from the poorest countries on the face of the earth an amount of money that exceeds what would be required to save 18 million lives by the year 2000, were that money available for the expenditure on human priorities. And if one forgave the debt—and in the case of Africa the overwhelming majority of the debt is held by governments, so it is possible to forgive it. If we forgive the debt, and turn the resources to child survival, then 18 million lives are saved. Can anyone explain to me what ethical considerations are involved if we lose 18 million lives for the sake of what amounts, in the minds of many African peoples and governments, to usury and extortion on the part of the creditor community?

Robert McNamara made a compelling speech on Africa on June 26, 1990 at the African Leadership Institute. He reproduced tables annotated by the World Bank indicating that from 1988 until the year 2000 a minimum of nine billion dollars would flow out of sub-Saharan Africa in debt-servicing obligations, unimpeded, *every year*. So what I'm talking about is not some kind of existential abstraction; it is a direct relationship. And those direct relationships are available in other equivalent areas. Were the developed world to overcome the stagnation in foreign aid—or, in the cases of my country and the United States, overcome actual *reductions* of foreign aid—it would be possible, with ease, to meet the additional seven to eight billion dollars annually that is required for sub-Saharan Africa.

To put it another way, were the developed world able to reduce the expenditures on defense and the military by a minute fraction, then there would be more than enough money to speak to the seven to eight billion dollars a year that is required to the end of the decade. Or, if the African governments themselves were able to restore their health and education budgets to levels of several years ago—levels eroded by the imposition of structural adjustment programs under the aegis of the World Bank and the International Monetary Fund—then again, the money would be available to save 18 million lives of children by the year 2000. What price, 18 million lives? It is to rage. It is to weep to see the emasculation of ethical considerations when attempting to assert the rights of the child.

This is particularly crucial for the continent of Africa because recent analyses suggest that the infectious reality of enveloping poverty is making the prognosis even worse for sub-Saharan Africa. On July 16, 1990, the World Bank devoted its world development report to poverty. A careful analysis of the figures provides some fascinating and profoundly distressing insights into the dilemma of the continent. What the Bank did was to look at the levels of poverty in 1985, and predict the levels of poverty in the year 2000 for the entire world. What it found is that the levels of poverty extant in the Middle East and North Africa in 1985 will be the same in the year 2000, and that the levels of poverty in what we once called Eastern Europe will also be similar in the year 2000 to what they were in 1985. But for Latin America and the

Caribbean, the levels of poverty will decline by 20% by the year 2000. For South Asia, poverty will decline by 30% by the year 2000. For India, by 40% by the year 2000. For East Asia, by 75% by the year 2000. For China, by 85% by the year 2000. *Only* in the case of sub-Saharan Africa will the levels of poverty *increase*: 180 million poor in 1985, 265 million poor predicted by the year 2000—an increase of virtually 50%.

UNICEF has a compelling maxim termed the principle of First Call. I don't see why I should adorn the principle when it is set out with such eloquence in the Summit Document that UNICEF has produced: "The foundation for the arch of protection which could now be constructed over the great majority of the world's children is a new priority, a *first call* for children. So important is this principle that in UNICEF's view it should increasingly come to influence the direction and nature of progress in all nations over the next decade and beyond."

In essence it implies that the growing minds and bodies of children should have a first call on our societies, and that children should be able to depend on the commitment in good times, and in bad. Translated into specifics, it means that whether a child survives or not, whether a child is well nourished or not, whether a child is immunized or not, whether a child has a school to go to or not, should *not* have to depend upon whether interest rates rise or fall, on whether commodity prices go up or down, on whether a particular political party is in power, on whether the economy has been well-managed or not, on whether a country is at war or not, or on any other trough or crest in the endless and inevitable undulations of political and economic life. That is to say, children have the first call on the resources of civilized society. Why? Because life is nothing without children.

One can resort to the cliches about children being our most precious asset. One can recite the components of vulnerability, of dependency and of protection that attend to the child. But there is, I think, a deeper, visceral, instinctive core that is universal. Simply put, there is no future without children, and everyone understands that.

This is a world where human rights are subject to every imaginable depredation, but the *reductio ad depravity* is the ethical shambles, the ethical squalor, when the rights of the child are abandoned. If the principles by which Albert Schweitzer lived were to govern the world in the next decade, it would be exhilarating to see what we could achieve for children. And I feel this morning, as I feel every day of my life, that the best is still possible if groups like this engage in advocacy that is tenacious, indefatigable and unrelenting.

Ethics
and
Human Rights

Lemke

Pawlowska

Hooks

Graesser

Bailey

Moderator's Introduction

By Antje Bultmann Lemke

This is the first symposium session in which the word "ethics" appears in the title. While Albert Schweitzer's ethic of Reverence for Life is the basis of his philosophy, the fact that the word has been chosen for this group of papers reflects the fact that there can be no realization of human rights without ethics.

To discuss this topic here is especially appropriate because it was at the United Nations that the first international Declaration of Human Rights was written and adopted under the strong and patient chairmanship of Eleanor Roosevelt, and then presented to the world in 1948. Thirty years later, in 1978, Hans Dietrich Genscher, as representative of the European Union, stated here at the United Nations,

> ... We are still far from translating the Declaration of Human Rights into reality all over the world. Nonetheless, we must not lose sight of the long-term trend of history. Since the proclamation of human rights, people all over the world have increasingly come to demand the realization of their rights. ... The rights of humans have become an international concern. Today they are one of the major issues in world politics.[1]

This is still so in 1990, and it is in the realization of the ideals expressed in the Declaration that Albert Schweitzer points the way.

Among the many topics Schweitzer addressed during his life, we find several references to our subject. In addition to various comments in his autobiography and his philosophical writings, two little known statements deal directly with the issue. The first is an essay on the relationship between the black and white races, published in 1925—long before any international declarations—and the second, a lecture with the title "Reverence for Life and Human Rights," delivered to the Association of Precision Workers in Paris in November 1959.

The international Declaration of Human Rights of the United Nations opens with the statement, "All humans are born free and equal in dignity and rights," a goal to which we all aspire. Schweitzer, in his first statement, discusses seven human rights in a very practical way, beginning with the right to shelter—to a place on this earth for every human being.[2] In the lecture of 1959 he outlines the principles upon which human rights must be based:

- Our basic rights are alike; only when we recognize and accept this can we develop a functioning society.
- We must formulate moral principles that go beyond geographic regions.
- We must respect each others lives.

Only if we have developed these three ethical principles as the basis for our actions and behavior, only when we have established trust, can we formulate and implement global declarations. Contemporary cliches often obscure real problems. For example, do we live in a global village, as Marshall McLuhan proclaimed in the 1950's, and as has often been quoted since? I do not think so. We have global networks that provide us with instant information. If we live in a village, we know our neighbors, we look into their eyes when we speak, we can touch each other, we share. But what we see on television rarely causes us to act, unless we are already motivated.

Just as slogans tend to cover up real issues, today's philosophical theories seem to move further and further away from reality. In this predicament, the speakers on our panel demonstrate how the ethical imperative of Albert Schweitzer can bring us closer to the realization of human rights in our society.

---

[1] Genscher, Hans Dietrich: Human Rights, an International Concern. *The Bulletin,* Press and Information Office of the Government of the Federal Republic of Germany. No. 10, vol. 5, p. 6, Bonn, 1978.

[2] Basic Human Rights, as Albert Schweitzer enumerates and discusses them in an essay, originally published as "The Relations of the White and Colored Races," *The Contemporary Review,* vol. 133, No. 475, pp. 65-70, London, January 1925: a. Right to shelter and residence (housing); b. Right to move freely; c. Right to the land and its resources, and the right to full enjoyment of both; d. Right to freedom of occupation and freedom of trade; e. Right to protection under the law; f. Right to live within a natural, national alliance; g. Right to education.

In the Absence of Love

By Pearl Bailey

Editor's Note: Pearl Bailey was an active member of the advisory committee of the International Albert Schweitzer Colloquium. She had already prepared her presentation for the colloquium when she died, suddenly, only days before the event. Her prepared text was read by Mpho Tutu, daughter of Archbishop Desmond Tutu.

How do we reach the goal of living in harmony, of having a relationship with our fellow man? Simple. Put love into action. Common sense, common decencies, mixed with the common knowledge that love is the major force, will put us all on the right path. Love without action is dead. Love is the major force behind all of our actions. When these traditional values of life (caring in all ways for our fellow man) are destroyed, then the foundation of mankind becomes quicksand and we all shall sink into nothingness. There will be no one left to throw the branches for us to hold on to; love will be our only reachable force. Human rights will be the only outlet to save ourselves.

Stereotyping humanity doesn't stop or start with the color of a man's skin, nor the heaviness of his purse, nor his religious beliefs. Why do we consistently mark minorities as the troublemakers of the world? Is it because they are the people who are so desperately in need? We have a tendency to label minorities as a retardation to humanity's growth. There is no intelligent reasoning as to why people are denied their rights.

Poverty and suffering have become universal, growing every day, and we are caught in the web of doom. We are caught in the web of the prophecies, man's destruction of man; doom and damnation will shower down upon us should we continue on this path we tread daily.

Man's greed for power over others is exceeding his ability to handle it. His lust for material gain, his selfishness, have taken hold. And so he ceases to function as a decent human being. We are becoming totally lost in *self*. We are no longer functioning in a Godlike way; no longer does our mirror of life show a Godlike image. What are these rights we dare to take from others?

Human rights is humanity acting upon and coming to grips with the challenges that face mankind each day: food, shelter, financial help. Take a long look at your measurements and you'll see man's material needs. They are the same as yours. The well of unanswered needs is a bottomless one. When you look into its waters you will see only a reflection of yourself. There but for you go I.

What if one devoted a portion of each day to trying to understand the problems facing us now? Sit quietly, meditate, move from yourself into another. The feeling will be tremendous. Picture yourself climbing a ladder of love, step by step from hunger and despair to a fulfilled life: food, shelter, the right to worship, a place for your family. As you climb, do not hesitate, nor look around to see who is following. It might be a lonely trip; then again, there might be a crowd. Regardless, keep on climbing. Fear not that you will lose your place on the earth because that place has already been recorded in the book of life. One asks, "Why should I be the one who reaches out? Can I, my country, my friends, alone help?" Yes! "Will my contribution be enough? Will it make a difference?" Yes!

No longer can people on this earth who are in need be ignored or destroyed because of personal conflicts. It is high time that we start to realize that we are dealing with flesh and blood. Who are we to decide what is enough for a man to have, or when it is necessary to give? Different people have different needs, but we all need. Allowing people to be swallowed up or pushed into corners to await our timing concerning their problems is inhumane. Too long have we practiced these methods; millions have been ignored, millions more have died.

Nations rest on their laurels as to what they have done, how wide they have opened their purses, how understanding they have been. The figures we face today on how many people have been mistreated in all manners are staggering, sickening and frightening. Did you help? Then help some more.

So many wait for others to start; so many wait to condemn others; so many wait so they can take over the weak; and many wait to figure out if it will be worthwhile to save these human beings.

Trying to bend others to our will is wrong. Preying upon their weaknesses—their hunger, their desperation—is wrong. If I feed you, clothe you, and loan you money, will you allow me to control you? What man, in a state of poverty and suffering, can avoid bending to his tormentors? But when he gets strong, he feels the need to break the chains that bind. Can you blame him? Once this occurs, this pitiful human being, who has had a crumb of bread, a mat for sleeping, a bit for his family, is called ungrateful. We tend to shackle helpless humans to ourselves and pull them into pits of deception. Release humanity by giving without restraints, loving others and encouraging them to make their way. Think back on how you made your way.

Injustice has played too large a role. It has been onstage too long and the right-thinking people of this world have tired of it. The applause for the hand-outs ceases. When does the permanent show go on?

We must attack immediately the lack of human rights; it is mushrooming as a nuclear blast. The ashes of "don't-carism" are being scattered all over the world. Impatience with these issues demands moral indignation. Words, promises are not enough. We need deeds—a continuity of deeds.

Let us unleash our God-given goodness upon the earth.

So many are thrown into a dither because they fear the equality of man. We will not all be alike, nor look alike, but we will all *be*. Alive and free. Mankind's skills should be exercised and encouraged for the good of his country, himself, and his family. Our society tends to segregate and categorize the true abilities of men according to what we think is best for this or that group.

Every person has, within him or her, a dream of excellence. Isolating a human being will not fulfill that dream. One of the greatest tragedies of our civilization is the emptying of the inner man, restructuring him into a vacuum, sucking him away from the stream of life so that he quickly becomes the dust from which he came, depleted of his spiritual resources. Lack of human rights in this world today is worsening; hope is fading for so many.

We must seek out those in need of love all over this universe. The dignity of man is a precious commodity. Spend it well.

Millions of people sleep with empty stomachs. Millions sleep under the stars. Try doing that, friends. It pains. Once tried, then go to the rooftops of our elegant homes, with full stomachs and heads full of knowledge, and speak with reverence and eloquence about human rights.

Wait for the echo of our pity, our lack of participation or compassion, our denials; then let our reasoning die down, and hear the dreadful silence of nothingness, the smell of death, the absence of love, saying to all, "I believe in human rights."

Man can act, if inspired. Man can respond, if called upon.

Let us inspire and respond before time is depleted. Millions die while we sit at the tables of discussion and say, "But there are so many." How right you are!

The Principle of Reverence for Life: Albert Schweitzer's Ethic for our Time

Translated by R.M. Gassen, T. Kaut, and A. B. Lemke

In 1963, two years before his death, Albert Schweitzer wrote these words to a friend in Strasbourg:

> My strategy is never to answer any attack regardless of its nature. This is my principle and I have adhered to it faithfully. No one will be able to fight silence in the long run. It is the invincible opponent. There is also no need for anybody to defend me. I am destined to follow my path without quarreling. It is my destiny to prepare the path for the spirit of Reverence for Life, which is also the spirit of peace. I am deeply moved to have been given this wonderful task, and the realization of this enables me to follow my road without inner struggle. A great, serene music resounds within me. That I live to see that the ethic of Reverence for Life begins to make its way into the world makes me impervious to any blame or attack.[1]

There are two reasons why I am moved by this late testimony. There is, first, the issue of defense: "There is no need for anybody to defend me." Surely not in the sense that Schweitzer's thoughts must be fitted with crutches to enable them to move about. The reflected and enacted thoughts of Albert Schweitzer need no apology. As an example of the complete identity of word and deed, they are the creditable testimony of a great humanitarian spirit against which charges can hardly be pressed.

But, although Schweitzer's ideas need no defense, they do need recollection. They merit our reexamination because they were far ahead of many concepts that today's philosophy only now—with difficulty and under pressure—has begun to realize: traditional anthropocentric ethical systems are not sufficiently comprehensive. Only morals enlarged to universal validity can be the "ethic of planetary responsibility"[2] required today.

Secondly, I am moved by Schweitzer's confession for its *confidence.*

Schweitzer died after a long and devoted life with the conviction that his ethic of Reverence for Life "would be recognized as self-evident [and] completely congruent with the nature of man,"[3] and that it would make its way into the world. He stated this conviction frequently[4], the last time only three months before his death.[5] Schweitzer sincerely hoped that his ethic of Reverence for Life might have the effect of salvation. He believed that we should become "different people, pious in an elementary, deep and vivid way,"[6] people "at home in the world and working in it in a higher way than was reflected by the prevailing [ethic]."[7]

It would seem to me, however, that here Schweitzer deluded himself. The facts point in the direction of a growing *lack* of Reverence for Life. The most significant evidence is the brutal recklessness with which the possibilities of modern advanced technology are enforced at the cost of non-human and even human lives. In order to finance their wealth and wasteful spending, the industrialized nations of the world have ceased to live on the interest from the rich capital of nature, and now live on the capital itself. Unrestricted exponential growth aimed at the final "conquest of scarcity"[8] may finally establish itself as a "suicide program" which we are powerless to reverse.[9] "We rejoice at the fire we have started, marvel at the beautiful flames, but still do not realize that they are about to eat us up," a contemporary critic wrote in the context of nuclear power plants.[10] We now pay a price for our energy consumption that no one can afford: the lethal waste we bequeath to uncountable generations to come. The penitential preachers, the admonitors with their skills of apocalyptic prognosis, had no major effect on the outbreak of the first oil crisis or on the disaster at Chernobyl. They have had little effect on the dying forests, the ozone hole, or the threatening collapse of world climate. "The shock of real and repeated catastrophes" had a stronger impact "than all the preaching of abstract forecasts by our scientists."[11] In this process it has become clear that the relationship of man and nature has entered a new phase:[12] man need no longer fear nature; it is nature that must tremble at the sight of man. He makes "the breaches" through which his "poison flows over the globe transforming the whole of nature into the sewage of man. We must protect the ocean from us, rather than protecting ourselves from the ocean. We have become more dangerous to nature than it has ever been to us."[13] Man, with his remarkable technology, has revealed himself as the number one planetary scoundrel."[14]

All this means, first, that our qualitatively changed technological potential, and the victories of civilization over nature, have become a universal danger. Man and nature are *both* subject to technology. Thus, the consequences of man's power, through technology, can mean his own defeat. Too great a "victory" threatens the victor.

Next, we must be concerned with finding the new ethic for this new situation. Hans Jonas, the Jewish philosopher living here in America, who studied philosophy with Heidegger and theology with Bultmann in Marburg, delivered what is presumably the most important contribution for this new ethic more than 10 years ago in his book, *The Imperative of Responsibility*. For this he was awarded the Peace Prize of the German Booktrade Association in Frankfort. Jonas proposes that life itself has been endangered through modern advanced technology. Using the principle of responsibility, he is in search of an ethic that protects not only human life, but *all* life. One would expect that Albert Schweitzer's ethic would come to his mind, but he mentions neither Schweitzer nor his ethic. Still, Jonas is very close to Schweitzer. He states that hitherto all ethical systems failed, that they offered no starting point for a responsibility toward the distant future. This, he says, is because all existing ethical systems are anthropocentric, and thus unfit for the new challenges of our time.

He fails to mention the only exception among the ethicists, Albert Schweitzer. An ethic expressing itself in its basic form as Reverence was introduced to the public for the first time in a sermon delivered in Strasbourg on February 16, 1919. It came to the public from the pulpit during a church service. The essence of this ethic is certainly not restricted to neighborly love of humans; it refers to all living beings, and it urges us toward a perception of the whole.

From the beginning, Albert Schweitzer dismissed the anthropocentric restriction of traditional ethics as "semi-ethics of the European philosophy." Against this, he promulgated "a deep and comprehensive ethic" to "create an ethical culture."[15] By this ethic, man's responsibility for his fellow humans and his solidarity with all creation are self-evident. At no time is it "meaningless to ask whether the condition of non-human nature, the biosphere as a whole and in its parts, now subject to our human powers, has become our trust and thus has a moral claim on us—not only for our sake but for its own, and in its own right."[16] Nature does exercise a moral claim on us in its own right, and not only for our own sake; naturally Schweitzer would have said for *its* own sake! "The love of creation, the reverence for all being, the sympathetic experience of life in all its forms, however unalike they may be to our form," is "the beginning and foundation of all morality."[17] It is therefore wrong, as Jonas claims, that no previous ethic—outside religion—had prepared us for a "role as trustees" toward non-human nature.[18] No previous ethic had prepared us for the "role as trustees" toward nature? This is false because Schweitzer's ethic of Reverence is intentionally not religious but, in his view, is founded on reason, on elementary thinking. Moreover, it is superior to other ethics insofar as the dualism between theory and practice, between ideas and interests, between academic creativity and non-academic activity, between ethics of knowledge and ethics of deed—this dualism has no room at all in Schweitzer's ethic.

The ethic of Reverence has no real need to experience the present anxiety, or, in Jonas's words, the "heuristics of fear,"[19] that impel us to look for new principles and for an ethic capable of fighting this threat. Schweitzer's ethic of Reverence has no need for anxiety in developing an ethic of responsibility necessary for our time. Reverence for Life is the basic moral principle for this responsibility. If man is touched by the ethic of Reverence for Life, Schweitzer claimed, he will not harm or destroy life unless it is absolutely necessary, and never out of thoughtlessness.

"In search of an ethic for the technological age": this is the sub-title of Hans Jonas's *The Imperative of Responsibility*. The search for an ethic for technological civilization need not—contrary to Jonas's assumption—begin in an "ethical vacuum."[20] It can be understood as a consistent development of Schweitzer's ethic of Reverence for Life, representing a reaction to technological development and the dangers resulting from it. The fact that Hans Jonas does not refer to Albert Schweitzer at all shows, indirectly, what an exception Schweitzer is among ethical thinkers. If one reads Albert Schweitzer, one soon realizes how close Schweitzer and Jonas are to each other. They have in common a concern not only for physical survival, but also for the integrity of the essence and the image of man. According to Jonas, "ethics must include both," i.e., the integrity of the essence and the image of man, and since one ethic must take care of both, it must go beyond prudence or ordinary knowledge; it must be based upon Reverence.[21] The "new kind of humility" to which, according to Jonas, we are forced by "the excess of our power,"[22] is in conformity with what had always been evident to Albert Schweitzer, the ethicist of Reverence.

Albert Schweitzer would feel validated that it is now said that the fundamental paradigm of morality is responsibility for the life I encounter, that is entrusted to me and is subject to my actions. Hans Jonas asked, "Can we possess an ethic—without returning to the category of the Sacred, which has been destroyed by the scholars of the Enlightenment—that is capable of mastering the extreme technological powers we possess, continually acquire and are all but forced to use?"[23] To this, Albert Schweitzer's answer could only be: certainly not! "Man is ethical only if life as such, that of plants and animals as well as of humans, is sacred to him, and if he devotes himself to life in need with a helping hand." [24]

Albert Schweitzer also clearly recognized the dynamics and the irresistibility of technological development—our predicament and, at the same time, our hope. It should be remembered that Schweitzer himself had realized the widespread endangerment of life caused by nuclear tests, and that he spoke passionately against them. In delivering his speeches against nuclear technology, he eventually enlarged his ethic of Reverence to an ethic of global responsibility. The "new territory,"[25] that we are about to enter is a no-man's-land. But the foundation for an ethic needed in our time has, in fact, been long since found. It is Reverence for Life. No "ethics of the new situation" can be so completely different that this foundation should prove irrelevant. The ethic of Reverence for Life is a compass that gives us an infallible guideline for man's course of action. We must make better use of this compass.

[1] H.W. Bahr. *Albert Schweitzer: Leben, Werk und Denken 1905-1965*. Mitgeteilt in seinen Briefen. Heidelberg, 1987: 322.

[2] Amery, C. *Das Ende Der Vorsehung: die Gnadenlosen Folgen Des Christentums*. Manburg, 1972 (rororo Sachbuch 6874): 231.

[3] *Albert Schweitzer: Gesamelte Werke, in funf Banden.* Muchen, 1974, Bd. 5: 170.

[4] H.W. Bahr. *Albert Schweitzer: Leben, Werk und Denken 1905-1965.* Mitgeteilt in seinen Briefen. Heidelberg, 1987: 206, 347.

[5] Ibid. 350 (Letter from June 9, 1964 to H. Mai).

[6] Ibid. 159.

[7] Ibid. 180.

[8] Spaemann, R. *Laudatio in: Friedenspreis Des Deutschen Buchhandels 1987.* Hans Jonas. *Ansprachen aus Anlass der Verleihung,* Frankfurt A. M., 1987: 21.

[9] Taylor, G.R. *Das Selbstmordprogramm.* Frankfurt A.M., 1971.

[10] Dahl, J. *Der wahre Preis des Stroms in: Natur. Das Umweltmagazin Nr.* July 1990: 98.

[11] Jonas, Hans. *Technik, Freiheit und Pflicht in: Friedenspreis Des Deutschen Buchhandels 1987. Ansprachen aus Anlass der Verleihung,* Frankfurt A. M., 1987: 45.

[12] Ibid. 39.

[13] Ibid. 37.

[14] Amery, C. Das Ende Der Vorsehung: die Gnadenlosen Folgen Des Christentums. Manburg, 1972 (rororo Sachbuch 6874): 233. Jonas, Hans. *Technik, Freiheit und Pflicht in: Friedenspreis Des Deutschen Buchhandels 1987. Ansprachen aus Anlass der Verleihung,* Frankfurt A. M., 1987: 42.

[15] H.W. Bahr. *Albert Schweitzer: Leben, Werk und Denken 1905-1965.* Mitgeteilt in seinen Briefen. Heidelberg, 1987: 328. The sentence: "All traditional ethics is anthropocentric." Jonas, Hans. *Das Prinzip Veranwortung.* Frankfurt A.M., 1979: 22.

[16] Jonas, Hans. Ibd. 29.

[17] H.W. Bahr. *Albert Schweitzer: Leben, Werk und Denken 1905-1965.* Mitgeteilt in seinen Briefen. Heidelberg, 1987: 125.

[18] Jonas, Hans. *Das Prinzip Veranwortung.* Frankfurt A.M., 1979: 29.

[19] Ibid. 63.

[20] Ibid. 7, 57.

[21] Ibid. 8.

[22] Ibid. 55.

[23] Ibid. 57.

[24] H.W. Bahr. *Albert Schweitzer: Leben, Werk und Denken 1905-1965.* Mitgeteilt in seinen Briefen. Heidelberg, 1987: 241.

[25] Jonas, Hans. *Das Prinzip Veranwortung.* Frankfurt A.M., 1979: 7.

Let Us Put Ethics and Human Rights Into Practice

By Benjamin Hooks

We are met at this time, and at this place, to pay tribute to the works of Albert Schweitzer, and to assess the relevance of his life and thought for our day and for the yet unfolded days that are ahead. What have we learned from this man of faith and reason, the minister and musician, the physician and philosopher? I believe that there is much that we can learn from this great humanitarian if we would hear and heed the wisdom of his teaching. Certainly he taught much, both by precept and example, about the topic of today's discussion: ethics and human rights.

Standard dictionaries yield definitions of ethics as "the discipline dealing with what is good and bad," or "a set of moral principles and values," or—another definition—"governing the principles of conduct." Ethics greatly concerned Dr. Schweitzer. In a lecture in 1921 he offered a definition of his own. "Ethics," he said, "is nothing else than Reverence for Life. Reverence for Life affords me my fundamental principle of morality: that good consists in maintaining, assisting and enhancing life. And that to destroy, to harm or to hinder life is evil." At another point in his life, he remarked, "The principle of Reverence for Life contains life's ethical affirmation." This is a notion with implications for the discussion of human rights.

In the United States and other Western democracies, with their free institutions, we have tended to think and speak of human rights largely with reference to societies other than our own. And then, most often with respect to societies that pursued a different ideology from ours. Within our own society, we may have spoken of civil rights, of civil liberties, but we have seen human rights as a problem for totalitarian, authoritarian societies, to note the distinction drawn by some political thinkers.

As a longtime participant in the American civil rights movement, I would say that I am in the human rights business. And since I represent the NAACP the National

Association for the Advancement of Colored People, the world's largest, oldest, most effective, most prestigious, most hated, most loved, most cussed and discussed human rights organization, I feel at home in this meeting. I recognize that some of my fellow citizens would not perceive any human rights problems in this country. Whether we speak of human rights or not, as an adherent of democratic ideals, I find great comfort in the governing principles set forth by Roy Wilkins, my predecessor as Executive Director of the NAACP. In one of his most memorable speeches, Mr. Wilkins declared of the American devotees of civil rights, "You are one of a little band, one of those in every clime, among people of every race under God's heaven. You believe that here on earth governments should believe in the goodness and wonder of individual souls. Woe unto that government, say you and say we all, that stomps on the believer and exalts the powers, the rulers of the darkness and exemplars of spiritual wickedness in high places."

One of the most difficult things in trying to uphold this concept of ethics and human rights is to speak of the government violating moral principles. I have been more cussed and discussed in the last few days because of statements that I've made in reference to [former Washington, D.C. mayor] Marion Barry. There will be some who say, "Why talk about a man who admitted to dope use, who is therefore possibly guilty of perjury?" But the principle that we must never forget is that we must never allow the government to use illegal or immoral means even to gain something which we think is right. If we do that, then we hark back to the days of Hitler. If we do that— if we let the government, with it's great power, overrule and overcome civil liberties, civil rights, human rights and ethics in order to obtain the prosecution of the wrongdoer—then we ourselves subscribe to the principle that ethics has no place, that it is more important to convict some guilty person than to uphold the rights of all. As long as I have breath in my body, no matter how you give me hell, I shall stand for the rights of people against the massive power of government to do wrong and be applauded.

That does not mean that we stand for the use of dope, nor that we believe that people ought to be wrongdoers. But, to paraphrase William Safire, who put it so well, "Perhaps in the history of this government, this is the first time we have used the methodology of the K.G.B. to entice someone, or bring someone into disrepute, through illegal means." And that to me is the whole question that confronts us today. (I would add that, if you have not read it, you should take the time one day to read Martin Luther King's "Letter from Birmingham Jail.")

For me, it is always difficult to talk about human rights and ethics in a vacuum. Being 65 years of age, and having spent more than half of my life fighting for rights that other Americans have taken for granted, it is never an easy subject. I think about the fact that, as Justice Marshall put it, as a black person I start out as three-fifths of a human—and then only for the purpose of representation in Congress. I think about women, who were not even given the right to vote until 1920. I think about white men, who arrogate to themselves all of the first fruits of everything. And now, as we come for reallocation and for justice, we hear cries—even from the White House—of reverse discrimination, quotas and other lying phrases that have no business in a country that believes in ethics and human rights. To paraphrase the writer Edward Markham, "They drew a circle and kept me out, but love drew a circle and brought them in."

And to those men and women and minorities who have suffered, it is a tribute to the greatness of America, as Bishop Tutu put it, that we have the Constitution on our side. I am happy to have been a participant in the nonviolent warfare to bring the ideals of Dr. Schweitzer to realization.

One day, as I thought about this subject while reading a book on slavery, a startling phrase occurred. It was that when slavery was first instituted it was *humanitarian* because prior to agricultural societies those who waged warfare killed all of their enemies. But when they found out that they could use them to raise potatoes and corn, wheat or whatever, then they imprisoned them. And this writer said that was an advance!

When I think of black people in this nation—244 years of labor from 1619 to 1863 without a payday; when I read with tears in my eyes the efforts of the Congress in the late 1860's and 1870's to redress that wrong; when I read the history of the 13th, 14th and 15th amendments which finally made the guarantees of the Constitution, and of the Bill of Rights, a part of my life; and when I see the assault today by the Supreme Court, I say that we need more meetings that will discuss ethics and human rights. Not dry legal theories, but ideas that will work in the great places where humans come together to build a society. To build the concept of freedom for everyone, and then deny free assembly, is the work of spiritual wickedness in high places. The denial of freedom of religion is the work of spiritual wickedness in high places. The denial of a free press is a denial of human rights and it is an elevation of spiritual wickedness in high places.

I read with interest—and this is a personal note—the long story on the Reverend Brother Al Sharpton in *The Wall Street Journal*. And I could not help but wonder that Ben Hooks leads an organization with 500,000 certified members and 2,200 chapters that, for 81 years, has been in the forefront of the freedom movement, and yet that greatly respected paper has never seen fit to do a profile either on the NAACP or on Ben Hooks. Instead, they do a story on the Reverend Sharpton, who said, "I can call a hundred people together every week." Yet I know preachers in this town who can call together 2,000 every Sunday. And they come enthusiastically. This makes one wonder what has happened to our concept of ethics and human rights.

And so, my brothers and my sisters, we live in a peculiar time. It appears that the winds of freedom are blowing in Eastern Europe, and even in South Africa. Although the problems that we are having in South Africa, even now, remind me that there are a great number of differences, and that the great spirit of hatred seems to permeate this world. Men and women are still fighting about ancient wrongs, so old that they don't even know what they started arguing about. And yet we have the words of Holy Writ, "And what doth God require of you, Man, but to love mercy, do justice and walk humbly before thy God."

I salute you for having this occasion. I'm glad to be a participant in it. And I hope that we can renew here a respect for human rights, and understand that respect for human rights is Reverence for Life. Injustices still persist in many places, but there is a spirit of freedom that is blowing across this world. And if we can catch that spirit, capitalize on it, work with it, find a methodology that can help us to ensure freedom for all women and all men of every color, creed and condition across this world, then our living shall not have been in vain.

There is a poem that I learned many years ago; I'm sure you've heard it. Some people associate it with war. I never associated this poem with war, but with the dream of peace, the dream of justice, and the dream of universal sisterhood and brotherhood:

"In Flanders fields the poppies blow / Between the crosses, row on row, / That mark our place; and in the sky / The larks, still bravely singing, fly / Scarce heard amid the guns below. / We are the Dead. Short days ago / We lived, felt dawn, saw sunset glow . . . To you from failing hands we throw / The torch . . . "

It seems to me that I can hear Dr. Schweitzer saying that. And others, who have passed on, who loved the cause of freedom.

"To you from failing hands we throw / The torch; be yours to hold it high. / If ye break faith with us who die / We shall not sleep, though poppies grow / In Flanders fields."

How Tolerant was Albert Schweitzer?

By Ija Pawlowska

It is taken for granted that we should be tolerant. Tolerance is certainly a high value: it protects the freedom of the human individual. This value can, however, come into conflict with some other value, and in a concrete situation we can feel that it would be morally wrong to exercise the virtue of tolerance.

My answer to the question I put in the title of my paper is: Albert Schweitzer was tolerant, but his tolerance was limited by his moral convictions. Here is an example of what he found absolutely intolerable. He wrote in 1935, "Today torture has been reestablished . . . in order to extract confessions from those accused. The sum total of misery thus caused every hour passes imagination." Schweitzer blamed the people who tolerated the practice: "To this renewal of torture, the Christianity of today offers no opposition even in words, much less in deeds . . . "

The idea of tolerance is closely connected with the idea of human rights. The Universal Declaration of Human Rights of 1948 proclaims for each individual a number of personal and civil liberties and aims to protect each individual against intolerance. According to the Declaration, everyone has the right to freedom of thought, of conscience and religion, of opinion and expression; no one may be compelled to belong to an association, and so on. However, we must be aware that the Declaration suggests also some limitations of tolerance, and it does it "for the purposes of securing due recognition and respect for the rights and freedoms of others and of meeting the just requirements of morality, public order and the general welfare in a democratic society." In particular, it is not permitted to engage in any activity aimed at the destruction of the many rights and freedoms set forth in the Declaration. So, as we can see, the idea of tolerance as the basis of individual freedom is not unbounded in the Declaration. But, on the other hand, there are also some definite limits to the ways intolerance can be practiced; Article 5 declares: "No one shall be subjected to torture or to cruel, inhuman or degrading treatment or punishment."

In my opinion Albert Schweitzer was as tolerant as is basically foreseen in the Universal Declaration of Human Rights.

Tolerance is concerned with beliefs and actions. The problem of tolerance occurs in situations when we are confronted with a lack of uniformity of beliefs or actions. Originally tolerance was equated with governmental non-interference in the sphere of religion. Nowadays tolerance is expected of the state, of social groups, of individuals. Contemporary conceptions broaden the idea of tolerance to include almost all spheres of life: religion, morality, politics, science, art, customs. I shall deal here only with religion and morality.

Until now I used the word "tolerance" without defining it. The word tolerance is being defined in many different ways. The two concepts that are most often found in books today, and that shall be useful in my investigations, are the concept of negative tolerance and the concept of positive tolerance. As I shall try to show, in each case "tolerance" requires a different form of behavior and does not necessarily refer to the same situations.

According to the negative concept, tolerance consists in a non-opposition to beliefs or actions considered as wrong. This definition takes into account two factors:
- What we tolerate is an opinion or action that we disapprove.
- We refrain from any correcting interference.

To be tolerant means to put up with something we dislike or even condemn. What we tolerate deviates from what we are convinced should be believed or done.

This concept of tolerance, which could be looked upon as the basic one, does not comply with some contemporary approaches to what tolerance really is or what it should be. The mere absence of interference in what we do not accept is treated by some thinkers as a too minimized and too one-sided interpretation of the idea of tolerance. Those authors expect something more than just non-opposition to somebody's otherness; they expect a favorable approval or even well-wishing support of the otherness. Professor R. M. Hare defines tolerance as readiness to respect other people's ideals as if they were one's own. Positive tolerance involves actually helping and encouraging the existing diversity. Thus, through respect and approval, and without interference, our attitude to somebody else's otherness changes from disapproval, to a positive evaluation and support.

The definition of this tolerance—positive tolerance—takes into account two factors:
- What we tolerate is an opinion or action that deviates from our own.
- We do not disapprove of this deviation.

Nowadays this kind of tolerance finds its expression mainly in the sphere of religion. For instance, if a Christian holds a negative attitude toward Buddhism, he will be accused of intolerance. "Not negative toleration but positive appreciation" was the postulate of S. Radhakrishnan, and what he referred to were religious creeds. Positive tolerance stresses the value of pluralism. To be *in*tolerant in the positive sense of "tolerance" it is enough to reject pluralism.

Albert Schweitzer was a Christian deeply connected with the Protestant religion, but he was truly tolerant in the positive sense toward other religions. This can be explained in the following way: religious dogmas and metaphysical conceptions were not, it seems to me, really important for him. Really important, always most important for him, were ethical beliefs. Here are some quotations to illustrate his attitude: "In religion there are two different currents: one free of dogma and one that is dogmatic . . . The religion free of dogma . . . is ethical, limits itself to the fundamental ethical verities . . . A time will come when religion and ethical thinking will unite." In Schweitzer's opinion, it is a mistake to demand from religion that it shall offer us "complete knowledge of the suprasensible. The deeper piety is, the humbler are its claims with regard to knowledge of the suprasensible."

For a few decades now, positive tolerance has been represented by some cultural anthropologists, adherents of so-called "cultural relativism." This is a result of a radical change in the attitude of Western people toward other cultures. The attitude that prevailed throughout centuries and is criticized nowadays as ethnocentrism consisted in comparing the tradition of other cultures with one's own and in condemning any dissimilarity as barbarous, pagan, primitive or abnormal. Intolerance led to extermination of entire tribes or even nations. Such extreme ethnocentrism grew to be quite unpopular and has gradually been superseded by extreme relativism, supported by the slogan of positive tolerance. According to cultural relativism there are no, and cannot be any, absolute standards for judging different moralities. Anything considered in a cultural tradition to be right, *is* right for the people of this tradition.

Of course, Schweitzer was not an ethical relativist. However, he did not simply compare the morality of other civilizations with our tradition, but with the moral pattern he worked out as the ethic of Reverence for Life. He used the same pattern to measure and to criticize the moral convictions and moral behavior prevalent in our own civilization. His postulate was "to measure all the principles, mental dispositions, and ideals which arise among us with the rule gauged for us by the absolute ethic of Reverence for Life." Schweitzer tended to accept as genuine only that which was "compatible with humanitarianism." He wrote, "The thinking man must . . . oppose all cruel customs no matter how deeply rooted in tradition."

Cultural relativists tend to divide mankind into many human groups or circles that are morally alien to one another and secluded to such an extent that the idea of a common standard cannot be applied to them. However, the ethics of Reverence for Life is intended as a *common* standard with *intercultural* validity to be used as a uniform measure of progress with reference to all human societies. Similar aspirations can be found among some earlier moralists and reformers; however, it was only in the middle of the 20th century that they were taken up, for the first time in the history of mankind, by an international institution supported by representatives of various cultures. The General Assembly of the United Nations proclaimed the Universal Declaration of Human Rights as a *common* standard of achievement for all peoples and nations irrespective of their cultural tradition. In opposition to the program of the cultural relativists, the United Nations aims to introduce *worldwide* the promotion of respect for, and observance of, the *same*—the *sane*—human rights and fundamental freedoms.

Let me emphasize once more that human rights are intended as fundamental and inalienable without distinction of race, color, sex, language, religion, political or other opinion, national or social origin, economic status or any other situation. Such an idea of moral universalism, Schweitzer represented also in his ethics. He wanted *all* people to enjoy as many benefits as possible, including personal freedom.

Now I would like to put a more general question: Can a person, especially a moralist, be tolerant of moral convictions and practices essentially different from those he himself holds and tries to follow? Can he be tolerant of convictions or practices that deviate from what he believes should be done if the matter is an important one? As an absolutist (and a moralist is, as such, an absolutist) he surely cannot be tolerant in the positive sense of the word.

In the negative sense, to be tolerant means to put up with something we disapprove, object to or condemn. When the case he faces is important, a moralist, I think, should at least not refrain from a *verbal* intervention, from criticizing what he regards as wrong—even if he does not have the power to impede or to stop what he condemns.

In this connection I want to present a third concept of tolerance that can be found in some books today. What tolerance in the third sense demands is not the absence of intervention, but desisting from the use of compulsion and any violent means against those whose moral convictions and practices differ from our own. What this tolerance does require is that we ought to find other instruments than violence or compulsion to get others to change their behavior. We may try to convince and persuade, offer advice and instruction—all carried out in a way that is inoffensive, not humiliating or otherwise hard to take. As long as it is all we do, we shall be acting tolerantly. It will not be the negative tolerance, because we take recourse to an active intervention, but tolerance in the third sense introduced here.

In the third sense, a moralist is tolerant as long as he promotes his moral program by argument, persuasion and living exemplification, and it must be stressed that the personal living exemplification was especially important in the case of Albert Schweitzer.

As a result of my consideration it can be said that Schweitzer was not tolerant in the sphere of morality either in the positive or in the negative sense of the word, but he represented tolerance in the third sense. In his text on Gandhi, Schweitzer said, "The most important thing is . . . that all worldly purposive actions should be undertaken with the greatest possible avoidance of violence . . . "

It is, unfortunately, not possible to be consequently tolerant—even in the third sense of the word—if we live in a society that secures the basic rights of its members.

Nowadays there is a tendency to differentiate between tolerance toward beliefs or actions, and tolerance in relation to people that stick to given beliefs and perform given actions. The social importance of this differentiation comes clearly to the foreground in cases when it is not tolerance that matters, but its lack—intolerance. The ways intolerance may be practiced *must have some limits*. These limits are described in the already quoted Article 5 of the Declaration of Human Rights: "No one should be subjected to torture or to cruel, inhuman or degrading treatment or punishment." Also, a criminal does not become worthless; he cannot be cast out of the family of man.

Schweitzer, as the great humanitarian, did not tolerate any kind of actions—including those considered to be lawful punishment—that were cruel, inhuman, or degrading to a person.

Ecology
and
the Environment

Ice   **Brabazon**   **Goodall**

Mittermeier   **Weaver**

Moderator's Introduction

By Jackson Lee Ice

In our first session we heard about arms reduction, the nuclear threat, and the problems of possible extinction of the human race. I call that "the quick bomb."

But there is another bomb, a slow bomb, that is just as destructive and just as deadly. Perhaps T.S. Eliot's prognostication is correct: that the world will end not with a bang, but with a whimper. The bomb that threatens us, that's slowly going off as we sit here, is our environmental crisis.

I feel as if I'm a citizen of a small town that's threatened by an oncoming flood, and that the way we are meeting this threat is by fixing a few leaking faucets in a few homes. I hope we can do more, and that we might begin by listening to our speakers attempt to make Albert Schweitzer's ideas relevant to this very important and very serious problem of our environment and ecology.

Reverence for Life: An Idea Whose Time is Coming

By James Brabazon

You will remember that in the Sixties a phenomenon occurred called "the love generation." Young people the world over subscribed to the theory that "love is all you need."

About that time, my wife and I were standing wearily in a line for a taxi after a long, hot journey, when several young people charged up the line and, with an arrogant show of physical superiority, took the next taxi that came. I asked my son, then about seventeen, what happened to the love generation. My son, a wisely skeptical lad, said, "Ah, that was last week. This week it's the hate generation."

This is a somewhat extreme example, but we all know that fashions in ideas change, sometimes as fast as fashions in clothes, sometimes more slowly. Look what has happened to Marxism in Eastern Europe.

Albert Schweitzer was not interested in fashion. He wanted to find a solid, permanent philosophical basis for ethical attitudes. And one would have thought that somewhere in the history of philosophy such a thing could be found. But search though he might, Schweitzer couldn't find one.

At the time when he was engaged in this search, the industrial revolution had recently given people of the West a new way of looking at things—the mechanical way, the technical way. The universe was not a divine creation after all; it was a machine that could be measured and analyzed and finally understood scientifically. And so with everything in it. The great theories of the 20th century all had this in common: they attempted a mechanical analysis of everything, including things that aren't machines. As Schweitzer pointed out at the time, the theory of natural selection suggests that nature itself is a great machine, working its way automatically toward ever more perfect specimens. Marx saw history as a machine for perfecting society.

Freud and other investigators of the psyche looked at the human personality as a machine that could be taken apart, analyzed, adjusted and improved. The whole study of economics presupposes that the market is some kind of machine that is capable of being understood and controlled, though the fate of numerous financiers and Chancellors of the Exchequer clearly demonstrates the opposite.

All through the 20th century we have been concerned to try to improve the system—the machine, the technology—and to find out more and more how things work, how the universe works; what makes it tick.

It was against the background of this buzzing scientific excitement, when all this was starting, that Schweitzer took the contrary view and said that to understand the relationship of human beings with the universe we must not look outward but inward. To look outward, he said, is a waste of time—nature is too vast and too complex ever to reveal all its secrets, or to enable us to comprehend our place in it.

The inward search into the psyche interested Schweitzer more. He even wrote a very learned psychiatric study of Jesus in answer to suggestions that Jesus was psychologically unbalanced. But his real search went deeper still, into the very core of human awareness.

And he was right. For looking inward we find a mystery as great as that of the origin of the universe. It's not just a question of how the idea of God entered into human consciousness—that can be accounted for by progression from the awe and dread of primitive races for uncontrollable forces of nature. The question really is, how did we develop the thing that we look into—the mysterious, unique *awareness* inside each one of us within which the concept of a God can arise?

"We look before and after, and pine for what is not." At a basic level, the ability to remember, to anticipate, to learn from the past and to plan for the future—the awareness of the possibilities of life, call it imagination—is what distinguishes us from other animals. It's been our most potent tool for surviving, for dominating the natural world. But there's much more that can't be accounted for merely by the need to survive. It is this "more" that we recognize as the most valuable thing about us.

Science has never yet explained what biological purpose is served by a Tchaikovsky piano concerto. Tchaikovsky looked inward and there he found marvels of beauty and intricacy. I only take Tchaikovsky because I happened to be listening to him as I was thinking about this talk. But there are all the other composers, all the other painters and writers—indeed all human beings, for at one time or another I believe that we all look within and find wonders of beauty and strange understanding, for which science could never possibly find a use.

And we find love.

We haven't begun properly to think about this incredible property of human life. Nature doesn't need love for the continuation of the species. Normal sexual attraction, call it lust, will handle that very well. So what function can it have, this thing called love? How does it come about that lust becomes transformed into something that can justly be called reverence—reverence for another person?

This is one of the many questions that forces us to go deeper than rational thought. Not to *abandon* rational thought, let it be noted. We aren't obliged to stop thinking when we fall in love. In fact, it's smart to think as hard as you can in that eventuality! But the fact itself is beyond thought; it transcends rationality. And this is what Schweitzer meant when he said that Reverence for Life was "a necessity for thinking." If you think hard enough about any of the real questions of life you come to the frontier, the boundary. And then you have to go on into realities that no longer accept rational explanations, *yet demand thought*. It is this that I want to emphasize.

Since Schweitzer formulated the phrase Reverence for Life, the 20th century has brought us face to face with what has been called "the death of nature." The mechanical way of thinking, combined with the destructive urges of the human race, together threaten to destroy the whole balance of the world.

In the same hundred years or so, every traditional value and ancient belief has been questioned and often abandoned. There seems to be no solid ethical gound left to cling to, no benchmark of quality of living to which we can appeal. A huge chasm has replaced our certainties.

Yet at the same time the barriers between the races are being broken down in a totally unprecedented way by travel, by radio and by television. As never before, we are aware of being one world—and a world with its mind marvelously concentrated by the threat of extinction.

And thus throughout the world there has been growing a sense of the sacredness of life. A hundred years ago, it was an unheard of notion that we should care for, and take responsibility for, every living creature. It would have been considered lunacy to suggest that we should preserve dangerous wild beasts or obscure varieties of spiders. Today it's almost taken for granted in large parts of the world. Perhaps we should remember that Schweitzer's was one of the first, lonely voices to say these things; but that by the end of his life his fame was such that practically every schoolchild in the world knew about his ideas and his example. I'm not suggesting that Schweitzer single-handedly changed the world's thinking. I am suggesting that his influence, even when we don't realize it, is everywhere.

But though we are beginning to understand the problems, and though the will to do something about them is growing, though we have immense new technical resources at our disposal to enable us to tackle them, the intellectual framework is lacking. The thing that we feel has no name, no one has thought through the implications. Overwhelmed by new experiences, new fears, new feelings, and without guidelines to make sense of them, we need desperately to reinforce our emotional responses with careful, unsentimental thinking to stop us from making new mistakes in our eagerness to put things right.

The old religions help, but all too often, while they comfort, they also divide. They are all bound up with tired old dogmas, and cannot cope with the flux of new situations and new ideas that confront us every day. Only a belief as simple, as fundamental and as universal as Reverence for Life has any hope of bringing and holding us together. The emotional impulses will die away into cynicism and "compassion fatigue," the love movements will change to hate movements, unless there is a clear understanding that we're *not* talking about empty idealism. We're talking about truth and reality.

Schweitzer got there a long time ago, and pointed the way. In the search for truth and reality, he preferred to avoid the use of the word "God," around which so much conflict and misunderstanding arise. And about the creator God he was not prepared to commit himself. Like an honest agnostic, he was not prepared to pin his faith to anything that he did not personally know. What he did know with absolute confidence, through experience and inward search, was his solidarity with other life. At a simple level, any small creature will teach the same lesson—a puppy, a lamb, an infant: they come into life with a fund of goodwill, a trust in other creatures and a delight in life that is only dimmed by the pressures of growing up. Schweitzer came to the conclusion that this consciousness of solidarity with all life is always there in every living being, and that any human being can use it who has the will to do so—though it is often so overlaid by bad experience, by hardened dogma and unchecked theory, or by the dismissive views of people who think it "adult" to be cynical, that it is no longer acknowledged, and scarcely even felt.

He thought this philosophy through and carefully analyzed its implications in his long book, *Civilization and Ethics*. He named his philosophy Reverence for Life, and he lived it day by day, in his African hospital and elsewhere, proving that it could work.

This awareness, this reverence for one's own life and that of others that Schweitzer found at the heart of Being, he was happy to call divine.

It is dangerous to call Reverence for Life a religion. But it is much more than an ethic. It is the drilling of a deep well in dry ground that brings to the surface a source of life and energy that was there all the while, untapped. Like a fertilizing river, it takes no account of frontiers. It crosses all barriers of nationality, political belief, religion, color and sex. In different circumstances it will give rise to different policies, but all will be humane, positive and fertile, so long as they remain true to the central impulse.

Thus it is not so much a policy in itself as a touchstone for all policies, and one to which all human beings can subscribe. Between peoples, it will seek always respect and peace. In agriculture, it will see that the resources of the world are husbanded, not exploited. In industry it will hold in mind that the purpose of manufacture is for the benefit of consumers, not simply to maximize profits. In finance, it will require that money is used for the enrichment of society through the fertilization of agriculture and industry, not for the making of sterile fortunes.

Reverence for Life is not just a simplistic statement that "love is all you need." It is not just a sentimental attachment to an idealized vision of "nature," for nature is indifferent to death and suffering. It is not simply a matter of self-preservation. It is not an impractical command for total vegetarianism, or a refusal to accept responsibility for making choices: for example, how do you keep a snake alive without killing mice? *You have to choose.*

It is the expression of what, deep down, human beings really are when they fulfill themselves. It is the liberation of the instinct in all of us to be an active and vital part of the living world, and a guide through the practical difficulties of how to do it. If we look deep enough within, thrusting aside the irrelevancies that we often cling to as principles, we can all find it.

Schweitzer knew that in the years ahead this belief, this underlying sense of what really matters, could and must be at the center of the human *mind* as well as the human heart.

Let us not be frightened by those who wave the word "realistic" in our faces, as though realism demands selfishness and excludes idealism. It doesn't. If this century has taught us one thing, it is that selfishness is not, in the long run, a realistic policy. Idealism is a matter of right thinking as well as right feeling.

Such a combination of heart and mind is still perhaps the most essential armor we can take with us into the future. The world is starting to catch up, but Schweitzer is still ahead of the game.

Some Lessons from Our Nearest Relatives

By Jane Goodall

It's a great honor for me to be here and bring the voice of the non-human animal kingdom very much into this distinguished meeting.

Albert Schweitzer said, "We need a boundless ethic that includes animals too," and a number of the speakers that I've heard today have indeed pointed to this philosophy in his writings and in his life. But I want to bring the non-human animal more directly in front of you, and consider very carefully this boundless ethic that includes animals too, over and above the reverence for life. I believe that Albert Schweitzer would have wanted somebody in this distinguished gathering to bring the non-human animals before you in this way. And I think I can serve the animal kingdom best by concentrating in this short time on the species that I know best, the chimpanzees, the animal that is today our closet living relative in the world.

To bring the voice of the chimpanzee directly into this meeting, let me give you a greeting from the chimpanzees of Gombe, if they were here. [She vocalizes, very strongly, a chimpanzee call.]

I think the thing that has been most striking during my thirty years at Gombe is that the more we learn about these extraordinary creatures, the more similarities strike us between their behavior and some of our own. Each chimpanzee has his or her unique personality; they're as different from one another as we are. Because I've been there for thirty years with my colleagues, we've had the opportunity to study, over this long period of time, how bonds develop between family members.

One such family was that of old Flo, her son Flint, and older daughter Fifi. Old Flo was still alive when Fifi, now thirteen years old, gave birth to her first known grandson, little Freud. (We thought Fifi was a sex maniac when she was an adolescent, so he had to be called Freud!) Some of you may remember that, when Flo became very old, she seemed to lack the strength to properly push her son, Flint, toward independence. He developed a strong bond with his mother and was abnormally dependent upon her. When she died, he was eight-and-a-half years old, and it seemed that he was unable to live without her. Flint was with her when she died. He showed symptoms of depression such as those seen in a human depressed child, and in this state of mourning, with his immune system weakened, fell sick and died about one month after losing his mother.

Fifi, left without a mother, showed the same affectionate, playful, tolerant maternal techniques as shown by her own mother. There is a very major difference between chimpanzee females in this regard. The young chimpanzee in nature has ample opportunity to play, to develop his muscles, to be full of fun and full of energy, just like many human children that we know. Objects are played with as toys. Youngsters are very inventive.

But it's not all play when you're growing up. As with human children, chimps have a great deal to learn. In different parts of Africa, for example, chimps have completely different tool-using cultures, and these are learned by the infants observing the pattern and then imitating and practicing. This probably is why, for five years, the young chim continues to nurse and remains very dependent on the mother.

The adults are extremely tolerant of youngsters, and for the first three or four years of life the adults of the community are affectionate toward the child, and he can get away with a great deal of mischievous behavior.

Weaning, the peak of which is around four to four-and-a-half, can be very traumatic. It seems to the youngster his world is coming to an end. The message of the mother when she calms his tantrums seems to be, "you can't have milk" or "you can't ride on my back, but I love you anyway."

Usually by the time the next child is born, the elder one is properly weaned and to some extent independent, but he remains closely emotionally bonded with his mother. She by no means turns him out to fend for himself. Because he remains with her and the new child, there develops a close bond between the growing siblings. It's very different for a second child in a chimpanzee family because there's always someone to play with. If his mother is not in the mood, then the elder brother or sister is likely to join him in his games. And if the mother dies, the elder sibling, whether female or male, will adopt the youngster and act *in loco parentis*.

The chimpanzees of Gombe are in a difficult situation now. It's a very tiny area, only about thirty square miles. There are three different social groups living within this area. They are now cut off from any interaction with other chimpanzees by the cultivation that has crept right up to the boundaries. It's almost like a glorified zoo. But, because for many years now we have employed local Tanzanians from the surrounding villages, not only to do menial tasks like trucking or anti-poaching, but to observe the chimps, follow them, record their behavior, and use sophisticated tape recorders and eight-millimeter video cameras—they have become passionately involved with the lives of these chimpanzees. They talk about them to the surrounding villagers; they bring their wives and children to see the chimps. We have absolutely no fear of poaching at Gombe.

However, the situation for the chimpanzees in the rest of Africa is grim, as it is for wildlife right across the continent, as it is (as we heard so eloquently this morning) for the children and the people living there, too. Chimpanzees are vanishing fast. They were present in 25 African countries. They're gone from four, they're on the verge of extinction in five others, and only in four countries in the central part of the range are they present in really large and healthy populations. They're disappearing because the forests are clear-cut for agriculture or human development. They're disappearing because timber merchants penetrate ever deeper into the forests, opening them up for hunters, and also taking with them the risk of disease. Chimpanzees are so close to ourselves they can be infected by all our human contagious diseases.

In some parts of Africa, chimpanzee meat is eaten. But chimps are hunted in all African countries, even if not for meat, for the sale of infants. A typical way of capturing the infant is to shoot the mother. You can image that this is not only cruel, but very wasteful. Often the mother is killed or mortally wounded and the infant who eludes capture, dies without her care. Even those infants taken alive from the mothers, even if they're not wounded—which they often are, with inefficient weapons being used—are quite likely to die during the horrifying journey from the dead or dying mother to the place where they are going to be sold.

These infants often have rough baskets woven around them. The infant will be tied at its wrists and ankles by rope or wire. It would make you sick to see the condition that some of them are in after a long journey through the forest. And when they arrive at a dealer camp, there's usually nobody there who understands their needs. A youngster needs his mother's milk and his mother's love. Instead, he's given a few bananas and a bowl of water, and there is nobody to reassure him in his time of need.

Some chimpanzees, quite a number in some countries, are bought by local people as pets. And, for awhile, they may live as part of the family. It may be an abnormal life, but if the new owner is understanding and compassionate, it may be quite a good life. But what happens when they get a bit older and potentially dangerous? They can bite. They are very destructive in a human house. They are highly intelligent; they can unlock locked cupboards and can very rapidly destroy a house and garden.

Some chimpanzees are smuggled out of Equatorial Africa into places like Spain, where photographers abuse them—they dress them up and use them to attract tourists. "Come and have your photo taken with a cute chimp as a memento of your holiday in sunny Spain!" They're beaten, they're very often sick, and they're drugged. We used to think they were drugged with things like Valium. It turns out, from some of the people who've been trying to confiscate and rescue them, that they will go through horrendous drug withdrawal symptoms because they are, in fact, on hard drugs!

Some youngsters end up in biomedical research. In this country, it's no longer legal for chimps to be imported from the wild, but they are bred in medical research units, where they sometimes wait in quarantine, living for months in small cages.

When they're older, they're put in slightly larger cages about the size of a normal podium. And sometimes these cages are put into something called an "isolet," a steel box with a small glass panel in the front. The only contact with the outside world is through an air vent, or when a technician opens the door to throw in some food, or clean them, or take their blood. They are alone at this stage, and they remain so for the next three to four years. While there are U.S. government standards on allowable cage sizes, it's perfectly legal at the moment in this country to keep adult chimpanzees, who may weigh up to 150 pounds, who may live 50 years, in cages measuring five feet by five feet and seven feet high.

Here, I think, is a sad commentary on Western science: scientists have been very eager to point to the close physiological similarity between humans and chimpanzees, and say that therefore they make good models for studying the nature, or searching for the cures, of human infectious diseases, such as hepatitis and AIDS with which most animals cannot be infected. Yet most people have been very reluctant to admit to the equally striking similarities between chimps and ourselves in the spheres of behavior, of the emotions and of the intellect.

And you know, looking back after thirty years, one sees that understanding chimpanzee nature does teach us a great deal more about human nature. It helps us to better see our own place in nature. And it's a little humbling; we're not quite as different as we used to think! We're not standing in isolated splendor on one side of an unbridgeable chasm. And in many ways, the chimpanzee serves as a bridge. It's very much easier for many people to make the intellectual leap over the supposed species gap between human and the rest of the animal kingdom, when they understand how like us the chimpanzee is.

It leads us, crossing the chimpanzee living bridge, to a new understanding of the rest of the animal kingdom; a new awareness, a new respect—a different kind of respect—for other life forms. It leads us into Albert Schweitzer's state of being reverent of all life.

Is there hope for the chimpanzees in the wild? Is there hope for the chimpanzees who have fallen captive into human hands? I believe there is. Many more will die. Many other animals will die, too, as the forest is relentlessly cleared. And the success or failure of human family planning programs will, to a very large extent, determine whether the rest of the animal kingdom in Africa will survive, the numbers it survives in, and the quality of life for the human beings living there today.

In African countries, we find a new awareness of conservation, of the value of conservation to the country. Yes, of course, it depends on the perception of the economic value, to the people and to their country. These are mostly very poor countries. We can't expect them to give up large tracts of land if they gain nothing from it. We must help, with offers of assistance in developing controlled tourism, for example, or of bringing in foreign exchange. We must bring aid from the developed world to the people living in these areas that are suitable for conservation. We must help rural development plans and provide medical care. We must help set up conservation education programs. It's no good imposing conservation on any country from outside. Countries are proud; it's got to be *their* conservation plan, for their own cultural heritage.

I want to tell you about Gregoire, a chimpanzee who was put into the Brazzaville Zoo in 1948. He's probably one of the oldest chimps anywhere in captivity. There was a rather meaningful incident which gives me great hope, great faith for the future of conservation, and ethical consideration of animals in Africa. I was standing in the Brazzaville Zoo, looking at Gregoire, with a high official of the Congolese government. He knew nothing about chimps, except what I'd just been telling him, and when he saw this emaciated Gregoire, who'd been in that cage alone since 1943, he was silent. And then he turned to me and said, "Jane, I think this is our Nelson Mandela." I thought that this was a very, very meaningful statement.

In zoos in this country and around the world, conditions for chimpanzees are continually improving. People are beginning to understand that chimpanzees are social creatures, with complex social lives, with long-term bonding between individuals. They are realizing that chimps cannot be treated as mere chattels, sent around the country, or left for their lifetimes in small, bleak metal cages. You've seen for yourselves the improvements in zoos, and it's worldwide. Even in some African zoos, when there's some money there, the animals are being kept in better and better conditions. In the labs, change is slow, but even in the worst labs there is talk of change. There are people being employed to enrich the lives of the chimpanzees imprisoned there. It's not nearly fast enough, but there is change.

There are people around the world, some wonderful people, helping in this fight to give animals a better chance of living decent lives when they're under the human yoke.

I want to end with one story. And probably some of you are familiar with this story because it happened in this country just a few weeks ago, in Detroit. A brand new chimpanzee exhibit opened for the first time this summer, supposedly the largest and best in North America. However, the chimpanzee group that was released into this closure was not well-advisedly chosen. They were chimps from around the world thrown rather higgledy piggledy (as far as I can tell) into a new social order. The first day, one chimpanzee jumped into the moat and drowned; chimpanzees cannot swim. And then, just a few weeks ago, one adult male got into conflict with the top-ranking male and was chased into the water. In his panic, he crossed the iron rail that is supposed to prevent drowning and he sank. There were a number of zoo personnel on the other side of the moat. They stood watching. There was also a visitor to the zoo, a truck driver, there with his family. He goes every year. Seeing the chimpanzee sinking for the third time in the murky water of the moat, he jumped in. He swam under the water until his hands felt the body of this chimpanzee. He managed to raise him to the surface, all 130 pounds of him, and he managed to lift him over the iron bar and push him up onto the shore of the island. All the time, the zoo staff were yelling at him to leave the chimp alone and get the hell back out of the water.

The director of the Jane Goodall Institute in this country, when he heard about it, called this man up. He said, "Rick, that was a dangerous thing you did. Why did you do it?" And Rick's answer was, "Well, I looked into his eyes, and it was like looking into the eyes of a man who was saying, 'Isn't there anybody who will help me?'"

And I think it's when all of us around the world hear that call for help, as Albert Schweitzer heard it all those years ago, and not only hear the call but respond to it, that's when the world can move ahead and come closer to a world in which there is a boundless ethic that includes animals too.

Biodiversity and the Inherent Value of Life:
A Modern Perspective
on Albert Schweitzer's Philosophy

By Russell A. Mittermeier

When I was asked to participate in this symposium on the Relevance of Albert Schweitzer at the Dawn of the 21st Century, I was very pleased and honored for several reasons. First, Dr. Schweitzer was one of my childhood inspirations, and his books were partly responsible for my lifelong interest in animals and the tropical rain forest. Second, although he loved all animals, he was, like me, and like Jane Goodall, very fond of our non-human primate relatives, the monkeys and apes. And as I understand it, he was especially partial to gorillas, and these were among the animals that most intrigued me as a child. Third, and certainly of great relevance to us at the dawn of the 21st century, was Albert Schweitzer's Reverence for Life.

This is what I would like to focus on in my presentation: the importance of this Reverence for Life. My interpretation of it will be in the context of today's many environmental challenges, and the need to appreciate, understand and conserve the fantastic diversity of life on our planet; something that I consider to be the single most pressing issue of our time. Other speakers have already made reference to this guiding philosophy of Dr. Schweitzer's life, but I'd like to read a brief passage that best encapsulates my interpretation:

"The deeper we look into nature, the more we recognize that it is full of life, and the more profoundly we know that all life is a secret, and that we are united with all life that is in nature. Man can no longer live for himself alone. We must realize that all life is valuable and that we are united to all life."

What I'd like to do today is to discuss how all life (what we in the conservation business refer to as "biological diversity") is currently facing its greatest crisis ever. As you all know, our global environment is under many different stresses, with problems like global warming, ozone layer depletion, pollution, erosion and toxic waste disposal becoming more and more serious every day. Not to mention the explosive growth of our own human population, which is at the root of many of our other problems.

However, I believe that there is one environmental issue that surpasses all others in long-term importance, and that is the loss of our planet's biological diversity. This diversity is our living natural resource space; our biological capital in the global bank, and its loss is an irreversible process. Although we can develop, or already have the technologies to combat other environmental problems, once a species of plant or animal goes extinct, it is gone forever. And we face over the next few decades a series of extinction spasms unlike anything since the loss of the dinosaurs, some 65 million years ago.

When we talk about conservation and biological diversity, worldwide, we're really talking about a major focus on the tropical regions of the world, and especially on the tropical rain forest, where Albert Schweitzer spent such a great part of his life. These forests are home to at least half, and possibly three-quarters or more of all species on earth, and they're disappearing far more rapidly than any other major biome. You've heard the estimates before: an area the size of a football field lost every second, 50 acres cut every minute, an area the size of the state of New York burned every year. Whatever the exact figures are, the fact is that these forests are disappearing very quickly and we really have only the next decade in which to act to ensure their survival.

Also at risk, but often overlooked, are the unique peoples of the rain forest, be they Yanomamo Indians from Brazil, Ibans from Borneo, or Huli wigmen from the highlands of New Guinea. The future of these people is closely linked to the forest, and their knowledge of the forest holds the key to using it without destroying it.

To me the main reason for conserving these forests and the teeming life within them is not economical or practical, but rather aesthetic and spiritual, as so clearly recognized by Albert Schweitzer. These forests are nature's works of art, the ultimate expression of the complexity and magnificence of life on earth, and they're worth saving for this reason alone. However, I think that most of us realize that in today's world, unfortunately, such arguments alone do not suffice. We have to look at the economic importance of these forests as well, and show how their conservation and rational utilization is at the very basis of sound economic development; that conservation and development must go hand in hand; or, better yet, that economic development in the tropics will quite simply be impossible without effective tropical forest conservation.

In developing strategies to conserve tropical rain forests, two things are of particular concern: the time frame in which to operate, and setting priorities for conservation action. Recent estimates indicate that we have already lost about two-thirds of the world's primary forest ecosystems. As if this weren't bad enough, look at the devastation that has already taken place in countries like the Philippines and

Madagascar where, in some places, up to 96% or 97% of all primary forests have already been destroyed, and where we have, at best, three to five years before all primary forest is gone. The bottom line: we need immediate action!

However at the same time, we have to set priorities. There are hundreds of different ecosystems out there, but some of them are simply more important and more in danger than others. We use a number of different approaches to set priorities, but the one that I'd like to highlight today, one that was first developed by my friend and colleague Dr. Norman Myers, one of the world's leading conservation thinkers, is what is called "the threatened hot-spots approach."

These hot spots are ten areas in the tropics that have exceptionally high diversity, very high endemism—that is, species that occur nowhere else—and that are under very serious threat at this time. These ten areas cover much less than one percent of the land surface of the planet, less than 4% of the remaining primary rain forest, and yet they contain more than a quarter, and perhaps as much as a third, of all tropical forest plants, and an even higher percentage of the world's animals. Two of these areas, the Atlantic forest region of Brazil and the island of Madagascar, are at the very top of the priority list. These are areas that I have spent a lot of time in for the past ten years, and they are indicative of what is happening to the hot spots in general.

The Atlantic Forest Region of Brazil is a unique series of ecosystems that's quite distinct from the much more extensive Amazonian forests to the northwest. It included some the most beautiful forests anywhere on our planet. However, this was the first part of Brazil to be colonized. It has developed into the agricultural and industrial center of this country, and it has within its borders two of the three largest cities in all of South America: Rio de Janeiro and Sao Paulo, the latter of which is one of the two largest cities on earth.

The result has been large scale forest destruction—especially over the last twenty to thirty years of rapid economic development—to make way for cattle pastures, plantations and industry. This is what's happening in the Atlantic Forest Region as a whole; we estimate that there is no more than one to five percent of the original forest cover remaining in this area. Needless to say, the plants and animals found in this region are suffering under such circumstances, and the primates that I have been working on there for the past twelve years are a very good example.

There are 22 different kinds of primates—monkeys—in this particular region. Seventeen of these are endemic, found nowhere else in the world, and fully 15 are already endangered. Two species of the Atlantic Forest stand out in particular: the muriqui which is the largest of the South American monkeys, and the golden lion tamarin, which is certainly one of the most spectacular of all mammals.

We've been using these animals as symbols, as flagship species to sell the whole issue of conservation within Brazil and internationally. We're working very closely with our Brazilian colleagues to spread the word about these animals and their rain forest ecosystems. We have major programs in place now, that began about ten years ago, that I think will be effective in ensuring the survival of these unique ecosystems for future generations.

Switching continents now, I'd like to take you to one of the world's most exotic places, the island of Madagascar. In moving to Madagascar we're not just moving to another part of the world, we're really taking a trip back in evolutionary time. Madagascar is a unique evolutionary experiment. It's a living laboratory that is unlike any place else on earth. Although it's located only about 400 kilometers off the east coast of Africa, it's been isolated for a very long time, perhaps as long as 160 million years, and most of the plant and animal species occurring in this unique country are found nowhere else in the world.

Although Madagascar is only about 40% again as large as the state of California, it's really a mini-continent in its own right, and has within its borders a wide range of ecosystems ranging from tropical rain forest, to dry baobab forests, to the unique spiny desert of Madagascar's south, all of this crowded into a relatively small area.

The most interesting and conspicuous group of animals in Madagascar, once again a group of primates but not monkeys or apes, are the lemurs. The Malagasy lemurs are a unique radiation that includes some thirty species ranging in size from the hairy-eared dwarf lemur (just rediscovered last year), one of the smallest living primates, up to lemurs as large as the indri, which looks like a cross between a teddy bear and a giant panda, and moves by bounding from tree to tree like an arboreal kangaroo.

One of the most striking things about the lemur radiation of Madagascar is that it's almost entirely endemic, almost entirely restricted to this island. Twenty-eight of the 30 species of lemurs in Madagascar are found only there, and about half of these are already endangered, some of them very seriously threatened with extinction and ranked among the most endangered primates on earth.

Lest anyone believe that extinctions are a figment of the conservationist's fertile imagination, he or she need only to look at what has already been lost in Madagascar since the arrival of our own species. Among the species that have disappeared are the elephant bird, the largest bird that ever lived. It stood about nine feet tall and had eggs that weighed twenty pounds.

We've also lost fully 14 species of lemurs, one-third of the existing lemur fauna at the time of man's arrival on this island, including a spectacular species, Megaladapis that looked like a huge koala and grew to be as large as a female gorilla.

In spite of all the problems in Madagascar, there's been a real turnaround in the past few years, and I think there's much cause for optimism in this country. The country itself is starting to realize, at the highest levels, how important its environment is for future generations. And the international community has started to take a major interest as well.

One of the things we're focusing on in Madagascar is conserving a core area of biological diversity, in the form of parks and reserves of various kinds, while at the same time taking into full consideration the needs of the local people living in the vicinity of these reserves. This is done on the premise that if we do not involve these people in all stages of conservation efforts, our efforts over the long-term will not be successful. These people have to benefit, short- and long-term, from conservation endeavors. Only in that way will we make conservation truly work.

I think it's happening now in Madagascar. I believe this is one of the hot spots where we really have a chance. And I am confident that we will be able to save a representative cross-section of Madagascar's unique fauna and flora for future generations.

So much for the priorities. How do we actually make conservation work? There's no one simple solution to achieving conservation in the tropics, and the fact is that we have to use a broad mix of conservation methods. They range from traditional endangered species conservation and national park establishment, to innovative new ways of integrating conservation and economic development. Furthermore, the mix has to be very carefully tailored to the needs and realities of each country and each particular place.

In my organization, Conservation International, we put a special emphasis on several different things. First of all, we believe very strongly in conservation science: that you can't do conservation without a very solid scientific underpinning. Second, we believe in the need to empower local people, and to work in full cooperation with institutions in the tropical countries to help them develop skills they need to carry the cause of conservation into the future. And third, we have an approach that we call "ecosystem conservation," which is an effort to integrate the needs of people with those of the wild species and the wild places upon which they depend for their survival, and to make this dynamic interface between conservation and economic development a positive force for the future.

Where do we stand right now? I think that we're really at the threshold of a new era in conservation in which this issue is finally taking its rightful place on the global stage. Just a few examples:

- There was a cover story last year in *Time* magazine on the destruction of the Amazon. This would have been unheard of just five years ago.

- In the Brazilian equivalent of *Time* magazine, *Veja*, we've had the same thing. Cover stories, key issues, all over the world.

- And a recent editorial in the magazine, *The Economist*, pointed out that, whereas defense was *the* global concern over the past forty years, in the future it's going to be the environment.

- Organizations that were never involved in conservation, and indeed were sometimes part of the problem, are starting to become partners in conservation endeavors. The World Bank, and some of the other multilateral development banks, although they still have a long way to go, are starting to become involved in conservation activities.

- Bilateral aid agencies, like USAID, are becoming active.

- The UN agencies, of course, are involved.

- And the European Community has done a lot as well.

I would like to note here that Japan now has the largest foreign aid packet of any country on earth, and we very much hope that Japan will become much more environmentally conscious, in the way they use this money, than they have been in the past.

Furthermore, with the incredible things that have been happening in Eastern Europe over the past year, I had hoped that the attention of the world would shift from the East-West conflict that has preoccupied us since World War II to what I see as the real challenge of the next century, the Temperate-Tropical dichotomy. It looks like we're yet again going to be sidetracked for a while by what's happening in the Middle East, but I hope very much that this will pass quickly.

Note here that I couched this dichotomy, not in the traditional North-South terminology, but rather Temperate-Tropical. It's in the tropical belt, sandwiched between the temperate parts of the world, that we really have to place major emphasis for the remainder of this decade and throughout the next century.

You are all familiar already with the great inequities in resource use, per capita income, food consumption, and so on that exist between the developed and the developing countries of the world. By early in the next century, 18 of the 20 largest cities, and 83 out of every 100 people on earth will be living in the developing world. But how many of you have thought of this in terms of tropical forest conservation?

Unlike our industrialized countries, the tropical nations depend very much on their natural resources for almost everything. And the increasing erosion of this resource base is a real recipe for disaster—disaster for each of the tropical countries, but also disaster in global geopolitical terms as well. As the natural resources of these countries are not conserved—and much of these resources are in the form of tropical forests—their increasing instability will become more and more a threat to world peace. I think that this issue of ecosecurity, ecologically-based security in global geopolitical issues, will become more and more of a concern in the future.

What is it going to take to conserve the world's biological diversity, this richness of other life on which we depend for our own survival? It's difficult to come up with the precise cost estimates, but I believe that we're going to have to make investments far beyond anything of the past. At the very least, we will have to approach this enormous challenge with the same levels of investment, energy and dedication as those with which we approached the space race in the 1960's. The difference is that, this time around, the stakes are much higher.

Just as an example of one of the things that's being done, Conservation International has a ten-year program entitled "The Rain Forest Imperative" which is focusing on some of these most endangered tropical forest ecosystems. We estimate that this will cost on the order of $250 million over the next ten years, and we think that this is just the tip of the iceberg. Many other organizations have similar strategies that they're starting to put into effect. It is going to cost a lot, but we don't have much choice.

The final message that I'd like to leave you with is that we have to be upbeat and optimistic about conservation. In spite of all the gloom and doom and endless reports of destruction, species loss and environmental disasters, there are real success stories out there. We have to learn from these, we have to build on them, and we have to work in the closest possible partnerships with our colleagues in the tropical countries to conserve our world's living biological resources. But perhaps, most important of all, we need to go back to basics, and look at Albert Schweitzer's philosophy of Reverence for Life more closely than ever before. Again, I quote, "We must realize that all life is valuable, and that we are united to all life. From this knowledge comes our spiritual relationship to the universe."

If we can fully understand both the value of our planet's living diversity, and our spiritual connection to all life, then I see no reason why we can't turn today's conservation dreams into tomorrow's conservation reality, and leave a living legacy to the 21st century of which Albert Schweitzer would truly be proud.

We Have But One Earth

By Dennis Weaver

I read someplace the other day that the earth is four-and-a-half billion years old, and that life, as we know it, has been on the earth for three-and-a-half billion years. And we humans have occupied it for only two to three million years, so we're really "new kids on the block." During that two to three million years, for the most part, we have lived in harmony with all other forms of life. That is, until recently.

During that period we had a few problems. We had the great flood, we had a few famines and a few small wars. But we didn't have the power to annihilate everything on this earth. We didn't have the power, in a moment of madness, to destroy the earth itself. But we now have that power. You see, we got smart—maybe too smart for our britches. We started going into this scientific age, and we started discovering things, and we got more and more intelligent, if you will. And as we did that, a strange thing happened. There was a tremendous shift in consciousness. And that collective shift in global consciousness brought us out of the dark ages into this period that we call scientific materialism. And that promised us so much. It promised us a more enlightened society, a more convenient and comfortable life. It promised us a life of abundance and pleasure and opportunity and, yes, freedom.

It also produced the industrial revolution through which more of that promise was fulfilled. Can you imagine what impact the industrial revolution has had on the lives of human beings? Just think back 200 years. Try to imagine yourself 200 years ago. No automobiles, no airplanes, no television, no radio, no satellites, no wonder drugs, no computers, copiers, compact discs—none of it!

Well, I want to tell you something. We didn't have something else 200 years ago. We didn't have acid rain, and we didn't have any greenhouse effect. And our ozone layer was not being threatened and destroyed. And there was no polluting and poisoning of our air, and of our water, and of the topsoil that gives us life.

You know, Emerson said a great thing in his essay, *Compensation*. He said, and he put it in language that we can all understand, "There is a tax on everything." He was just talking about the law of cause and effect: wherever there is an action, there's a reaction. There's a price that we pay for everything that we do. And there was a huge price, *a huge price*, paid for the industrial revolution.

One person I read said that the industrial revolution was like a huge party. We were on a tremendous binge and now we're being handed the bill. And the bill, for our actions, and for the industrial revolution, is simply that we may not now have an environment in which we can live and continue life as we know it. That's the bill!

We're supposed to be the intelligent species . . . right? Well, we're destroying the very thing that allows us to live. Do you realize that we have painted ourselves into a corner? The very thing that we must have to live—the society we have made—is now destroying us. Think about it! How smart can we get? Well, we might just contemplate a little bit, and try to figure out how we got into this mess.

There are a lot of short-term reasons, but we've got to look to the very core, to the very basic reasons of why we are in the situation that we're in. And Dr. Schweitzer hit it right on the head. He said that the disastrous feature of our civilization is that it is far more developed materially than spiritually. Its balance is disturbed. Now come the facts that summon us to reflect. They tell us in harsh language that a civilization that develops only on the material side, and not in the spirit, heads for disaster.

The obvious question, if you accept that, is how do we develop spiritually? How do we bring balance to our own lives, and thus to the world? It must start with the individual. I would suggest that we have to wipe away the ignorance of our own spiritual nature. We must once again know in a real, and specific, and personal way, that consciousness of the spirit within, that perfect reflection of the creator that each one of us carries deep within. That is the soul itself. We must know that.

I'm not talking about joining a temple, or joining a church. I'm talking about going into that silence within, and making that inner search. We've been searching *outward*. And we've ignored the inner search. That's where we must turn if we are to develop spiritually. We must know that eternal part of us, and know through our own experience that the nature of that consciousness is peace, is love, is joy. And it is only when we feel that inwardly that we will reflect it outwardly.

One great Indian sage said, "Utopia must first spring in the private bosom before it can flower into civic virtue." Coming out of the dark ages, the collective shift in consciousness did not include spiritual development. We were simply too busy with the material side. Now, let me try to make this point: a shift in consciousness is triggered by a major change in the way that we perceive the world around us. And as we perceive the world around us, our attitude changes about that world, and our behavior changes.

So with the development of science, our perceived reality of the world around us changed. We were no longer seeking salvation through blind faith, but through advanced technology and science. We thought of ourselves not as an integral part of nature and this earth, but separate from it, even superior to it.

Sir Francis Bacon probably expressed the prevailing attitude more clearly than anyone else when he said, "The purpose of science is to wrest nature to the ground, and to extort from her her secrets for the benefit of man." To wrest nature to the ground? You can't wrest nature to the ground! Nature will win every time!

As Emerson said, "Nature will always try to balance itself." And in that effort to balance itself, if that means wiping away a few million human beings, that's the law. You cannot win. You can only get in harmony with nature. You can't defeat nature! Can you feel the violence in that statement? "Wrest nature to the ground . . . extort from her . . . " A violent attitude leads to violent actions. And so we have violently raped, pillaged, plundered and abused this earth.

And the earth has reacted violently to that abuse, as any living organism will do. What would you do if somebody poured acid rain on your head? What would you do if somebody poisoned the blood that runs through your veins, that makes it possible for you to live? What would you do if somebody took a knife and stabbed you? You would react! You would try to get away from it; you would try to do away with that which is trying to destroy you. Because a desire to survive is natural to every living organism.

As Dr. Schweitzer said, "I am life that wills to live, in the midst of life that wills to live." Every living organism has a will to live. And every living organism will fight to survive. And the earth is no different. It is a living organism that is fighting to survive. And that fight is reflected in increased, and greater, and more intense tornadoes, earthquakes, floods, droughts, famines, etc.

We can compare the earth to our body. Both have immune systems. Both are living organisms. Both can take a great deal of abuse. Our bodies can become very diseased, yet heal themselves if we give them an opportunity. But there comes a point when our immune systems can no longer handle the abuse, can no longer handle the disease, and what happens? We die.

Since the earth is a living organism, the question is, when do we cross the threshold? How much abuse can we perpetuate upon this earth before it is irreversible, and the earth dies?

I don't want to sound like there's no hope, because there certainly is hope. We have the power of choice, and we have the power of will, and we have the power to take the actions that we need to take to create a world of peace, productivity and justice. The real hope, in my view, for our future, is that we are right now in another great shift in collective consciousness!

And whether we survive this very exciting but very dangerous period that we're going through depends upon how quickly and how effectively we make that shift. Very simply put, that shift is the *knowing* that we are *not* separated, that we are all connected, that all life is one.

Take the smallest atom or the greatest cosmic body: there is not one single thing in this entire creation that isn't somehow connected to, identified with, dependent upon, or effected by some other one thing. Nothing lives in isolation. And because of that connection—and this is the higher truth that we're coming to—we share each other's pain and we share each other's joy. We are simply strands in the great web of life. And whatever we do to the web, we do to ourselves. That's the understanding that we must come to.

We share each other's pain and we share each other's joy. If we could just *understand* that! If we could understand that, and allow our behavior and our actions to spring from that understanding, it would change the world overnight! It would solve every problem that we have.

Knowing that simple truth, we would certainly restore our precious environment. One of the great challenges for the '90's—and we've got a number of them— is how to convert from a wartime to a peacetime economy. That's one; here's another: how do we marry our ecology to our economy? Because we need both of them.

That's why I coined the word *ecolonomics*. If we don't have an economy that's sustainable, that allows us to bring food to our table, that's chaos. And if we don't have a sustainable environment in which to live, that also is chaos.

There are many challenges. But the number one challenge, as far as I'm concerned, is: are we going to be able to take this tremendous technology and use it, not to demean life, not to destroy life, but rather to enhance life, and to glorify life? That's the challenge!

It was Bertrand Russell who said, "The only thing that will redeem mankind is cooperation." And I sincerely believe that there is tremendous power in cooperation. We don't do anything by ourselves, but there is nothing we can't do if we do it together. That's the power of cooperation working.

The only thing that will redeem mankind is cooperation. I believe that because I believe it's a form of love. And there is no power in this universe greater than the power of love. One great saint said, "We will either learn to love one another, or perish."

Cooperation and love are no longer an *option*. If we are to survive, they are an absolute necessity. We will either arrive at the point of understanding that we are individualized parts of one whole, or we will continue to believe that we are separate, that we can somehow bring benefit to ourselves while we damage others. And if we hold to the latter belief, we will absolutely destroy ourselves.

Greed, as you know, is probably the genesis of most of our problems. It must be replaced by love and understanding and cooperation.

Now I know some people say, "That is so idealistic; that is so naive." They say, "Greed will never go; we can't get rid of greed because people are motivated by self-interest." We *are* motivated by self-interest. We all want to be happy; we all want to be fulfilled. That's the nature of each and every one of us. And what we do, we do in order to try to achieve that. So we are motivated by self-interest, but there is a large self-interest and there is a small self-interest. The small self-interest thinks only of the little self; the large self-interest thinks of the larger self, and understands that our individual happiness must include the happiness of others.

So, self-interest will always be with us, but that does not necessarily mean that greed will always be with us. Greed will go when our own self-interest demands it, when we realize that it is a blight on our own happiness.

The collective shift in consciousness that we're talking about, that we're entering into, is not something that can be legislated—no more than goodwill, brotherhood or love—but it will happen as the hearts of people change. That's the key. The hearts of people must change. It doesn't make any difference how many laws we make. We've got laws against murder; we've got laws against stealing; we've got laws at an international level that say it's time for no more wars. But we don't pay attention to them. Why? Because the heart is not right.

The hearts of people must be changed. I say this often and people say, "Yes, the hearts must be changed. I'm going to go and change all those hearts out there." What a futile effort! There's only one heart we can do anything about, and that's our own. When we change our own hearts, and we start acting from kindness, from understanding, from tolerance, then we become a real force. Because if you change your own heart, you will change thousands. That's the key, and that's what's happening. That's what's creating this great shift in consciousness.

So our contribution to the dawning of the new world, one of peace, one of productivity, one of justice, and one in harmony with all of nature, is to light our own candle. When we do that, then others can light theirs off of ours. That's when we really become effective.

Hold this picture in your mind: candle after candle after candle being lit, just sweeping across the world like a great prairie fire, pushed forward by the winds of love, burning away all bigotry, all prejudice, all fear, all greed, all anger, all hate, and leaving in its path nothing but peace, and cooperation, and a world that lives in harmony and justice, safe and secure. Hold that vision!

A lot of people ask me, "What are you doing in Hollywood to try to move this thing along?"

I'm very proud of the fact that we have an organization there, the Earth Communications Office, made up of writers, directors, producers, actors—all creative people. And our role is to communicate. So you will find, in television shows and movies, a lot of environmental problems being brought up to increase public awareness. When people from CBS came to me and asked me if I'd do another Sam McCloud, I said I would if we could update it. I didn't want him to still be in New York City trying to learn the ways of big city cops. I said, let's upgrade him; let's make him significant to today's world; let's make him a Senator—Junior Senator from New Mexico, elected on an environmental platform.. They agreed; they felt that was harmless. So I suggested that at the beginning of the show we should find Sam McCloud on the Senate floor fighting for his environmental bill, trying to save the ozone layer. They agreed again.

But the only problem was, nobody could write the speech. That shows you where their consciousness was. So I said, let me take a whack at it, and they did. I was absolutely amazed when they accepted it. As it begins, Sam McCloud comes to the Senate floor and there's a guy there filibustering, trying to shoot down his bill. And Sam grabs the microphone, and this is what he says:

> Mr. Chairman! I want to apologize to the Senator from Wisconsin, but
> I've got something to say about this ecology bill, and I'm not leaving until

it's been said. Now, we've all come to Washington for various reasons. I came with a feeling for the land.

You see, the most important issue that we all face—whether we're old, young, black, white, male, female, rich, poor, Democrat or Republican—is how do we save our common home, this earth, on which we all live? Because if we don't figure out a way to do that, the deficit, sure as hell, is not going to matter. Neither will unemployment, taxes, interest rates, homelessness, hunger, human rights, drugs—none of them! Because, my friends, if we don't save the ozone layer, the simple truth is that we won't have to worry about those things because none of us will be here.

Now to jog your memory, the ozone layer is that protective shield around the earth that filters out the deadly ultraviolet rays of the sun and allows life to exist on this planet. And we're destroying it; there's a hole in it at the South Pole that's bigger than the whole United States, and it's growing as we speak. And when it's gone—that's it! I mean, you can't run down to the general mercantile and buy another one!

Now, we are the only animal that destroys the very thing that allows us to live. How stupid can we get? I'm not here to paint a picture of gloom and doom, but to point out the magnitude of our opportunities—they're tremendous. There is nothing, absolutely nothing, that we can't do if we use our time, our energy, and our money in a sensible way—if we get our priorities straight. Do you realize that within the last six hours of this debate we have spent more money on weapons than we have on the environment in the last ten years! I don't know how you feel about that, but to me that's the tail wagging the dog.

This is no time for finger-pointing, because we're all guilty! We've all been part of the problem, and we better damn well all come together and be part of the solution! And I'm talking about labor, industry, management, scientists, artists, professionals and politicians. We have simply never faced a situation like this before. The question is: do we survive? Do we continue, as a species, or do we just join the dinosaurs and vanish?

Because, my friends, when that ozone layer goes, we all go. Position won't save you. Power won't save you. Money won't save you. Are we no better than a bunch of lemmings running headlong toward mass suicide? Good heavens, we're supposed to be the intelligent species. We are the custodians of this planet, the protectors of all living things on it, and we've failed. We've failed to protect them because there is something more to protection than guns and bullets.

Now we, in this assembly, have a special responsibility . . . there's only a hundred of us, and we're speaking for over 250 million.

That is an awesome responsibility, so when you vote, just ask yourself, "Am I voting for life? Am I voting so that my children, their children and theirs can also know life . . . can walk upon this earth . . . and breathe its pure sweet air . . . and feel love for each other?"

You make no mistake about it, this wondrous Mother Earth that nourishes each one of us is hurting. She's in pain. She may be dying. She needs our help. She needs your vote. It's your choice. It's your choice! You sleep well tonight.

By the way, I might add, the bill was passed.

In Conclusion

Muller

In Conclusion
Some Highlights, Suggestions and Hopes

By Robert Muller

I have spent a total of 43 years in this house. It is one of my three great loves: my wife and my family; the United Nations and the University for Peace; and also the country that comes closest to my heart because it fulfills practically all my ideals, Costa Rica, a country that has had the courage to demilitarize itself. The country is a paragon of democracy, which has known no war, in the middle of Central America, where so many wars have been going on.

Two days ago I was sitting with guerillas from El Salvador, at the University for Peace, and listening to how they want to imitate the example of Costa Rica in asking for the demilitarization of El Salvador. We are working on the creation of a zone of peace for all of Central America, on the pattern of Costa Rica.

My life has been full of dreams, and most of them have seen results. As I listened to Oscar Arias, I had another dream. It is that I would be buried in Costa Rica. Because there is no military, I think it is the only country where I will really rest in peace! And I will have a condition that if ever Costa Rica remilitarizes, my bones may be transferred to another country that is demilitarized.

It's difficult to conclude this colloquium after having heard such a wonderful conclusion from Dennis Weaver. But let me say that, a few years ago at the University for Peace, we had a thought that we should honor great peacemakers of history. So we came up with the idea that, every year, in cooperation with the United Nations Peace Studies Unit, we would have a conference.

We had the first conference on Teilhard de Chardin, the second on Thomas Paine, and the third—last year—on Tolstoy. And this year, Albert Schweitzer— which is the best of all. As we are speaking about Teilhard de Chardin, I must mention that there is a friend of mine here who is the secretary of the Association of Alsatian Writers. She has given me a wonderful paper detailing the lives and thinking of Teilhard de Chardin and Albert Schweitzer. It shows exactly how two people started with theology, then each became a scientist—one a paleontologist, the other a medical doctor. And as a result, they were able to develop philosophies of life that were constantly transcending themselves into the universe, and in time. I had never thought of this parallel between the two thinkers.

The idea of these symposia—and now we have four of them—is to provide building blocks for the next millenium. We will have more, one every year. We will have to plan one on Gandhi, and we will also have to find a good woman peacemaker, like Jane Addams or Eleanor Roosevelt. We will have to select them in such a way that by the year 2000, out of these symposia, we will have the new philosophy of which Rhena Schweitzer is dreaming. Because the present world has no philosophy; it is at a loss. But we can put everything together in order to provide a new philosophy for our future. And we have ten years for this.

In creating the symbol of this conference, Dr. [David] Miller has done the right thing by drawing these circles of human rights, health care, the environment and arms reduction—the four topic areas of this colloquium—with peace, the result, in the middle. And I hope that, for all four sectors of the conference, we will have an inventory of the practical proposals that were made.

For example, Homer Jack's proposal that the United Nations should be transformed into a world government. Or the marvelous proposal that the UN Volunteer Program should be a World Volunteer Corps. And even the proposals of some of the young people here who have asked me to work toward the preparation of a youth conference to precede the Earth Summit in Brazil in 1992, so that the Earth Summit will include the perceptions of youth. From now on I will make sure that every world conference is preceded by a conference of young people in order to get the perceptions of the "new cosmic units," the new generation that will have the future as its charge. It is a wonderful idea—we cannot leave youth out of this. They are the new perceptive units of this planet, of humanity and of the cosmos.

I would like to say to them, as Linus Pauling said, that it is a wonderful period in which to live. Or as Charles Dickens says at the beginning of *A Tale of Two Cities*: "It was the best of times, it was the worst of times . . . " We have heard here about the worst of times and we have heard about the best of times. Personally, I am always inclined to see, first, the good side, in order to derive hope. As we're sitting here, we are a completely new species. We have extended our eyesight into the infinitely large and into the infinitely small through radiotelescopes, electron microscopes, and so on. Our hearing goes around the planet. Our hands are transformed and multiplied by machines, our legs by cars and airplanes. Our brains, our memories, are transformed by computers. It is incredible how we have become a totally new species.

Now, is this wrong? Perhaps this is the evolution that was bound to happen; namely, that one species would get out of its lower phases and begin to identify itself with the totality of the universe. We began to do it with our senses. Now we are beginning to do it with our brains; we are increasing our brains. The dinosaurs are said to have died because they had little brains, and we must hasten to get big brains to survive in our current world situation. But we are only beginning this evolution. We have not yet tried with our hearts. We have no global hearts, no global spirituality.

I have lived this in the United Nations because here you have to take care of the living first—to avoid war, to deal with hunger, to deal with sickness. If you started a debate on God in the United Nations they would say, "Look, for heaven's sake, the first rule is 'don't let the children die.'" You heard this from Stephen Lewis this morning. We can discuss God later on . . . but it will come. We have had very deeply spiritual Secretary-Generals: Dag Hammerskjold said that we needed a spiritual renaissance. The Buddhist, U Thant, who was for me a father and a master, believed that spirituality was the core of life; he couldn't understand the separation of spirituality from everyday life. For him, every moment was a spiritual moment; spirituality was the greatest virtue of all. So this is beginning in the United Nations too. But we are still in the kindergarten of this global age.

You have a wonderful target: the year 2000. We have 10 years. And we have to prepare for a new millenium. I think this is the first time ever, in the entire human history, that humanity goes toward a new millenium having the necessary knowledge about this planet. We didn't even know that the planet was round 500 years ago. We thought that the sun was turning around the planet; then came the Copernican revolution. We didn't even discover America until less than 500 years ago.

When I joined the United Nations, we had no global data at all. This is why we have the world population explosion—we had no good idea of how many people lived on this planet until 1951. Today we have the data. We have the knowledge—from outer space to the atom—so that you, the young people, have all the files on this planet, on humanity, on any group you can imagine—religions, cultures, nations, professions: you have it all. But it is up to you now, out of these files, to fashion the future and to create these harmonies: with the planet, among humans, with the heavens, with the past and with the future. This is a great, beautiful job you have!

As Linus Pauling said, we should be happy to be alive today. He also said that he likes to solve problems. I once suggested, to the Union of International Associations in Brussels, the preparation of a repertory of world problems. After two or three years, they published an enormous book: there are 11,704 world problems!* So don't tell me that you, the young people, are bored because there is nothing to do.

There is a lot to do, and you have beautiful dates in front of you. You have 1992, which is the beginning of a new European Community. Can you imagine, I come from a border from which I could see the Germans on the other side. And I saw the birds and the moon—they didn't see any border. My father was once a French soldier, then a German soldier. My grandfather had five nationalities without leaving his village! When I was in the French Underground, I swore that I was going to work for peace all my life, and I have done this. Today my passport says "European Community" and, underneath, it says "France." And in 1992, if I'm still alive, I will go to my home town to see that damned border dismantled.

*The 1991 edition of the *Encyclopedia of World Problems and Human Potential* lists 13,167 world problems.

When you live in a region that has two cultures, the only option is to find humanity. Germans, French—why aren't we all *human*? Why don't we have a European Community, one Europe? It was another man from my region, Robert Schuman, who did it. Today we have a united Europe, and now the Eastern European countries are rushing in to join too. So this dream of Robert Schuman has been fulfilled. He was a deeply spiritual person, like Albert Schweitzer. I am a member of a committee created in Alsace-Lorraine for the canonization of Robert Schuman, a political man, whom we want to see declared a saint. Of course the Vatican says, "How many miracles does he have?" And my answer is, "Look, he didn't heal one person, he healed millions, because now there's no war anymore between France and Germany." This, that seemed impossible, has become possible.

So 1992 is going to be an important year toward the creation of a World Community. Mr. Bush has hastened to propose an American Community, from Alaska to Tierra del Fuego—wonderful, at long last they are discovering Latin America, and giving hope to the idea of making this a beautiful bi-continent.

In 1992 we're going to have the Earth Summit in Brazil, 20 years after the first one, to see where we went wrong and what has been improved. This is a tremendously important conference and I'm glad that you, the young people, approached me immediately to find out how you can participate. The 500th anniversary of the discovery of the New World will also be in 1992. Now we have to find another New World, beyond the Americas and all the nations of today. This will be your world, which you have to discover for tomorrow.

In 1993, we're going to have the 100th anniversary of the first World Congress of Religions, for which Vivekananda came from India to Chicago and said that all religions have the same denominator. The anniversary conference is going to be one of the greatest gatherings of religions on this earth. And there we shall ask the religions, "What are the cosmic laws? What are the divine laws?" And I hope that they will adopt as one basic law, which I'm sure Albert Schweitzer would approve, "Thou shall not kill, not even in the name of a nation or a religion."

For 1994, the United Nations has declared the International Year of the Family. How wonderful! At long last, the family, the basic social unit. I hope that the Pope will come here, visit the United Nations, say what he has to say about the family, and also give his spiritual view of the next millenium.

In 1995, we're going to have the 50th anniversary of the United Nations. In San Francisco, they're already preparing this anniversary. All the people who have been declared Peace Messengers by the United Nations are going to work together to make 1995 the Year of the Family of Nations.

We have to culminate all this, in the year 2000, with a Celebration of Life. Between now and then, put your thinking caps on. Visualize the type of New World that you want. Let the best thinkers from all professions prepare for a new millenium, a completely different millenium, so that we can rid the world of those wrong priorities that have been underlined here in this conference. Let's stop spending our resources on things that are obsolete. According to Soviet scientists, we have 16 years to live before this planet is unlivable. The Worldwatch Institute in Washington gives us 40 years. Lovejoy in England gives us 20 years.

I would like to see a conference of scientists at which we would ask every scientific group, "How long do you give us to live?" My daughter, who is an ocean biologist, believes that we are finished because business and government do not listen to the scientists. Let's get all the scientists to tell us, more or less, "How many years do you give us?" How long do they give us until the ozonosphere is depleted, the seas and ocean dead, the oxygen and water in shortage, and the radiation of the planet lethal?

Let's change things so that we don't say, "What is good for General Motors is good for the world," but, instead, "What is good for the world is good for General Motors."

To you young people, these would be my recommendations based on my experience: Try the impossible! I have become an expert on impossibilities. I'm not interested in something that's possible, because everybody does it. I'll go for the impossible. When we began the United Nations, they told us that decolonization would take 150 years. It was done in 40 years. They tell us today that demilitarization is not possible. Demilitarization will take place too. They tell us that nuclear disarmament is impossible. Well, a number of these arms have already been suppressed, and now they are going to discover that they can do this, too. All these things are possible. To save the environment is also possible. Select what you find to be the most impossible.

Ask yourself, "What were you born for? Why do you live?" And make this your purpose in life, work on it every day, work on it a hundred times. We have learned here, in this conference, that the individual has tremendous power. Albert Schweitzer said, "It is only the people who can push governments." As the Chinese and as Gandhi say, "If the people lead, the governments will follow." This is absolutely right; this is true democracy. So look into yourself. Don't imitate fashions of the day. Don't listen to the advertisements. Don't listen to the statement that you have no power. Five billion people on this planet have the most incredible power on earth! So exercise this power, express your views, write letters, "network" day and night. Prepare yourself in the evening: when I go to bed, I put a thought in my mind so that the following morning, when I wake up, I write an essay on it. My subconscious goes to work on it. I don't like to lose my time during the night.

The basic secret is to see yourself as part of the total universe, the total cosmos, the total stream of time. We are cosmic units, allowed to live an increasing number of years.

But we are born, and we die, like the cells in the body. We are all part of the body of humanity, and humanity is part of the body of this earth, and this earth is a part of the cosmos. It is a marvelous earth on which life has been born in such multiple forms. This is why the religious tell us that we are children of God, that we are divine. Yes, we are divine; we are cosmic units. And if you look at this from the point of view of the universe, of space and of time, then everything falls into place. Then national policies and idiosyncrasies look completely irrelevant. Live in the universe, and live in eternity. Each of you is a conjunction of the universe and of time. You are extremely precious, unique, never-to-be-repeated entities in the entire history of the universe. There will never be exactly the same being again. You are great! When we celebrate great people, we make a little bit of a mistake. We should not say only that they are great, because *each of us* is great.

Sometimes people say, "Oh, Mr. Muller, you are special because you have been with the United Nations for forty years." That's a mistake. Because what we want to do is make out of each child a great cosmic unit, fulfill each child's potential. We have to validate people, and not invalidate them as we so often do. Albert Schweitzer told us that we must validate life to its fullest, not diminish it.

So here, young people, take this into your hands. You can change the world. You must change the world. You have to try the impossible. Demilitarize this place! There is no reason why this beautiful earth should float in the universe with atomic explosives in its flesh, in its waters, in its atmosphere. They even want to put nuclear devices into the stars! You cannot accept this; it is your responsibility *not* to accept it. They can tell you any stories; don't accept them until they change their stories! Please, I beg you.

Notes

The following are excerpts of panel and audience discussion that followed the speakers' presentations.

Session I

Following Leaning's presentation:

Marshall (moderator): Dr. Leaning has spoken for the physicians. Mr. Safronchuk has emphasized the need for every group and individual to take a position on these urgent problems. What do the other panelists, from the religious, scientific and communication communities have to add to this?

Jack: I worked for 13 years with a small group, The World Conference on Religion and Peace. Dr. Schweitzer was one of the originators of the idea that world religious leaders, from Buddhism to Zoroastrianism, should come together to oppose the nuclear tests of that time. Back in the 1960's, a diverse group did come together; it was the beginning of many such efforts by religious leaders, but tragically, we never really did enough.

Pauling: I believe that it is the duty of every human being to do something toward eliminating the barbarism of war in the modern world. And I feel that scientists have a special obligation. Almost all problems today have some scientific basis. Consequently, scientists need to help educate their fellow citizens to reach the right decisions.

Following Cousins's presentation:

Marshall (moderator): This has been a valuable testament of Norman's thinking and personal experience with Dr. Schweitzer. Perhaps other members of the panel would care to give their sense of what Schweitzer's yardstick might have been on this current crisis in the Persian Gulf.

Jack: If Schweitzer were here he might ask President Arias to walk up that corridor to the Security Council, that even now is in critical deliberation, to plead with the U.S. representative to adhere to those virtues of which he has just spoken—humility, prudence and patience.

Marshall (moderator): Dr. Leaning, you did not personally know Albert Schweitzer, but you do know his ideas, which you have so superbly presented here. What additional thoughts do you have on how he might have reacted to the Gulf crisis?

Leaning: Two things strike me in his writings. One is his enduring sense of empathy for individuals who suffer. He would be appalled at the potential for the use of weapons of mass destruction in the Gulf, and what that would mean to the captive populations in that region. The other is that, although he thought that non-citizens usually should not offer advice in the affairs of other nations, in this current crisis he might be impelled to urge otherwise. Since it is unclear how the citizens of Iraq, as well as of other nations, might express their opinions in this matter, Schweitzer might consider it reasonable to urge individuals everywhere, as citizens of the world, to call upon world government in a way that Mr. Cousins has described. And because the United Nations is the court of last resort, it might rise to the occasion, as we have always hoped that it would do.

Marshall (moderator): When I visited Schweitzer at Lambarene, he told me of his search, during the early years of World War II, for an underlying ethical basis for civilization. This search, in part impelled by the tragic breakdown of civilization as manifested in that war, made Schweitzer recognize the crucial need for us to expand our boundaries of who constitutes "us" as opposed to "them" beyond the usual considerations of family, tribe and nation—and even beyond humankind. It culminated in a memorable moment of intuition when, in 1915, the words Reverence for Life flashed upon his mind, and became thereafter the foundation of his ethical philosophy.

Discussion then followed on the dangers of nuclear tests and nuclear war, on the post-World War II Cold War, and on Schweitzer's somewhat ambiguous and puzzling position on colonialism and African independence.

Rhena Schweitzer Miller (from the audience): Although my father did believe that independence for Gabon came before the people were prepared for it, once the decision for independence was made, he respected it absolutely. My father always respected the Gabonese government and the government has always supported the Schweitzer hospital.

Session II

Following Dr. Foege's presentation, a questioner from the audience expressed concern over excessive energy consumption, its consequent greenhouse effect, and the aggravation of the Third World's problem of undernutrition by increases in the rate of population growth.

Foege: The biggest problem we now face is the environment-population nexus. New information, less than two years old, tells us that the best single predictor of what will happen to birth rates in a country turns out to be what has happened to mortality rates in the previous ten years. Conversely, we can also show that money spent on family planning reduces not only birth rates, but also infant mortality rates. The two are synergistic.

Sidel: I'd like to add to that. Most of the per capita energy use on this planet, most of the contribution to the greenhouse effect, comes from the industrialized countries—from the rich countries of the world. Little contribution comes from the poor peoples of the earth. Therefore, if we are concerned about this threat to the planet, as we must be, we must ask what the rich nations are doing to cut down on their contribution to the greenhouse effect. We must make sure that they do not use the greenhouse effect as an excuse to deny poor countries the ability to industrialize and to reach the standard of living and health that the industrialized nations have reached. We must in no way blame the victims of the world, the poor nations, for the greenhouse effect. And we must look to the industrialized nations, to ourselves, for the solution to that problem.

Karefa-Smart: I am now devoting nearly all of my time to the problem of excessive population growth. In 1974, when an international conference on world population growth was called in Bucharest, only one African country—Ghana—subscribed to the view that an over-rapid increase in population was a great deterring factor to social and economic development. All the other African countries said, "This is not our concern; this is a concern of the Western world." Ten years later, at another world population conference in Mexico City, all the African countries—*all* of them—said with one voice, "We have come to the conclusion that everything we do in our economic planning for the betterment of our people will be brought to naught if we continue to allow our population to grow at a rate that is *so* much faster than the rate of our economic development and productive capacity." A good example is my own country, Sierra Leone. When I was a child, our population was only a million and a quarter. Today our population is over four million. What has it meant to the health of the men, women and children of my country? It has meant that today we are actually farther behind in our progress toward health, universal primary education and a better life for all our people than we were when I was a child.

But I want to say something else that some of you may not like. The United States government, which took the leadership role in the efforts to make the whole world aware of the dangers of too rapid population growth, today is taking a *back seat* in these efforts. The present administration actually refuses to contribute to the one United Nations organization whose principal task is to assist developing countries in dealing with this critical problem . . . If I were an American, I would be ashamed.

A member of the audience expressed appreciation that Dr. Jampolsky and Ms. Cirincione brought the life of Christ and the love of God into the colloquium, saying, "I don't think one can even begin talking about world peace unless one talks about Christ."

Sidel: Some of us are atheists. Some of us are Jewish. Some of us are Moslems. We too feel we can do our bit for the world and for peace. While we deeply respect the religious basis that led Schweitzer to do what he did, and that may have led people on this panel and in this audience to do what they have done, that is not the only route to individual self-fulfillment and, more importantly, that is not the only route to make certain that one's life is a life of justice, of service to others, and of dedication to world justice and peace.

Another audience member, saying he worked for New York City's Board of Education Health Department, asked why, if doses of vaccine cost only about five cents, his clinic does not have sufficient vaccine for all the people it serves.

Foege: The answer to this question is that you can't buy vaccine for that price in this country. It's a long, complicated story. Much of the vaccine is produced by expensive research in this country, so the companies here have to charge a great deal. More important, we are a nation that sues, and so the price of the vaccine must include the cost of legal action taken because of ill effects from the vaccine. On the world market, we don't have to worry about this, so immunization is a great bargain in the developing world.

Dr. Miller ended the discussion after this session with a reminder of Schweitzer's maxim: "Ethical belief should lead to action."

Session III

An audience member from the New York City-based Albert Schweitzer Fellowship expressed concern over the current situation in South Africa. He had lived in Soweto, had visited other South African cities, and stated his opinion that many recent events in that country "are symptoms, not causes." He expressed his belief that a deeper set of causes is operative, and asked Miss Tutu to comment on "the current tragedy in South Africa."

Tutu: The root causes that allow the current violence all stem from the apartheid system. There are divisions and power struggles among leaders of the tribal "homelands," of the African National Congress, and of the South African government and its police forces, that often seem to harm rather than protect black citizens. There's a tremendous level of rage, even among those who do not support one side or another, but are simply victims of the violence. To hope that these problems, having existed for decades, can all be resolved in a matter of months is unrealistic. We have great faith in humankind and great hopes for the triumph of the human spirit, . . . but we must also recognize the fallibility of human beings.

From the audience, an Alsatian medical doctor and theologian who is director of the Gabonese foundation directly responsible for the Lambarene hospital: An African at Lambarene said to me, "Tell those people in New York that Dr. Schweitzer's greatest achievement is shown by this: Today a man from the tribe of the *Fang* and a man from the tribe of the *Galoa* will together carry a sick patient who is from the tribe of the *Eshira.*" The Albert Schweitzer Hospital was created as a sign of reparation for the terrible harm that white people have done to black people. It is based on the help of those who have been freed from illness and pain, and now feel the responsibility of helping others. It operates primarily as a private hospital, but with significant support from the government of the Gabon. It is our responsibility to see that it can continue to operate.

Another audience member, from the former German Democratic Republic, citing the right of peoples to self-determination, and the power of informed public opinion to effect change, said that, to his mind, the recent peaceful revolution in his country has demonstrated the validity of Dr. Schweitzer's ideals.

Following Graesser's presentation:

Lemke (moderator): I just want to add that, like many European philosophers, Hans Jonas had overlooked Schweitzer. After *The Imperative of Responsibility* was published in 1979, I pointed this out to Jonas and he agreed that he should have considered Schweitzer. When he received the Peace Prize of the German Booktrade in Frankfurt in 1987, a prize that Schweitzer had received in 1951, Jonas gave a lecture, *Technology, Freedom and Duty,* that was like a sequel to Schweitzer's earlier lecture, *What Kind of Peace for Mankind.* Both underlined the interdependence of all life, and expressed the hope that human reason can pave the road to peace.

Following Pawlowska's presentation:

From a retired United Nations staff member in the audience: The Universal Declaration of Human Rights includes the right to work without being discriminated against because of race, sex or age. Why then, does the UN, which was instrumental in developing the Declaration, not respect this right with regard to its own staff members? What can be done to remedy this?

Muller: Efforts are being made to raise the UN retirement age above its present level of 60 years. However, these efforts have thus far been unsuccessful because of pressure from the African, Asian and other new member states, and from the younger people eager to be hired by the UN to replace the "old timers."

From the audience, a student with the youth organization, Global Kids: What you have been saying here yesterday and today concerns us very much. I hope an effort will be made to encourage the participation of young people, along with adults, in meetings like this.

An official from a European Albert Schweitzer organization volunteered: I would like to add something to what Professors Graesser and Pawlowska have said. Reverence for Life did not just fall full-blown from heaven. At least seven disciplines have come together in its creation, including philosophy, theology and nature. It is a synthesis of the ethic of devotion and the ethic of self-perfection. It does not make life any easier; rather, it makes it much more difficult. But at its uppermost level, we reach what Pearl Bailey so movingly spoke about: love. Only when we are capable of Reverence can we truly love.

In conclusion:

Lemke (moderator): Throughout this session, people whose words and deeds are embedded in the ethics of human rights have come to mind. Among those I think of are Vaclev Havel; Teddy Kollek, the mayor of Jerusalem; and Oscar Arias, who was with us yesterday. Their examples show us the power of ethical principles, which our speakers have so eloquently conveyed. As we think of our own spheres of responsibility, we should remember that Albert Schweitzer never intended to provide solutions for specific problems of today or tomorrow. But his philosophy and his example are a foundation on which we can build, with strength and conviction, as we face the future.

Session IV

Following Goodall's presentation:

Ice (moderator): The insight of Schweitzer's ecological ethic certainly fits the problems we now face with the animal world and with the environment. But when it was first proposed it was considered quite novel to think that we should expand our moral precepts and behavior beyond the realm of the human, and actually consider behaving morally toward other animals, even including birds and insects. This ethic was proposed long before most of us had even heard the word *ecology*.

After Dr. Mittermeier's presentation, a member of the audience asked if the goal of ecology is to preserve as much diversity as possible in order to understand it scientifically, or to restore a natural order that has been harmed or destroyed by human activities.

Mittermeier: First and foremost is the recognition of our complete interconnectedness with all other life. Here in the Western world we tend, in our culture and our religions, to take an anthropocentric view of the world. We must understand, much more fully, how closely connected we are to everything that's out there. It can be couched in economic terms, in aesthetic terms, or in spiritual terms, but the fact is this: the connectedness is there, it is real. Only when we really begin to recognize it, can we begin to come to grips with these issues that we are facing.

A student in the audience asked Dr. Goodall's opinion on research laboratories using animals for testing.

Goodall: This is an extremely difficult question. First, I don't feel that it was right, at the outset, for us in our arrogance to assume that we could simply use other animals in the ways we have. Second, we must realize that, if we are to continue to use animals in research, we should not use them in testing commercial products. I am not a medical doctor, and I cannot speak on how necessary or useful it may be to use animals in medical research. But we must face the fact that, from the perspective of the animals, they are being tortured. I can only fight for improved conditions, the use of fewer animals, and closer scrutiny of procedures.

Another student in the audience asked about the involvement of local people in the areas where animal conservation work is being done. He felt it would be wrong if local people were not actively included.

Mittermeier: You've hit the nail on the head. In fact, one cannot effectively do conservation work in these countries without full involvement of local people. One learns this very quickly. One can pass all the laws in the capital cities, but nothing good will happen locally unless local people see how they, as well as the animals, will benefit. Frequently indigenous groups like the Amazonian Indians have much better knowledge of how to coexist with their environment than any of us. They can teach us a great deal more than we can teach them. This is something that conservationists are finally realizing.

From the audience: How do we prevent mismanagement and bloated government from co-opting this very important work?

Mittermeier: I think one of the key interests over the next decade will be to bridge the gap between high-level interests (governments, the military) with lots of talk (environmental workshops, green summits, and so on). We must translate this combination into effective action on the ground, quickly and with minimal bureaucracy. This is an important role that the non-governmental community can take. I think it's great that governments are working on environmental concerns, but they are indeed bloated and slow-moving. The non-governmental community, both in tropical countries and in this country, must increasingly play a key role to facilitate the kind of rapid and effective action we need.

Weaver: It's always a problem of how to get the government to do what we want it to do. Gandhi said, "When the people lead, the leaders will follow." We must become informed. We must understand the power of the pen. And we must get involved.

Following Mr. Weaver's presentation, a member of the audience observed that much of what had been said during the Colloquium seemed to reflect the Buddha's feeling of compassion for all being. Responding from the audience, Dr. David Miller noted Schweitzer's testimony that, in 1915, on the Ogowe River, when the phrase "Reverence for Life" flashed upon Schweitzer's mind, there also appeared in his mind's eye the vision, not of Jesus, but of Buddha.

About the Contributors

Oscar Arias Sanchez capped a career in government service with his election as president of the Republic of Costa Rica in 1986. In 1987, he received the Nobel Peace Prize for his efforts at constructing a just peace in Central America. President Arias is also the 1989 recipient of the Albert Schweitzer Humanitarian Award. He serves on the Board of Advisors of the Albert Schweitzer Institute.

Pearl Bailey, entertainer and author, was special advisor to the U. S. Mission to the United Nations and a public delegate in the United Nations General Assembly. She was a recipient of the Presidential Medal of Freedom, the Golda Meir Fellowship Award, and numerous other honors. She died in 1990.

James Brabazon has been an actor and playwright, and is now a television director/producer and an author. His 1976 biography, *Albert Schweitzer,* has been acclaimed as "the best balanced and most complete account of Schweitzer yet to appear." Brabazon also completed work in the former Soviet Union on a film about Stalinist labor camps.

Diane Cirincione, a cross-cultural facilitator for attitudinal healing, lecturer, author and entrepreneur, is the founder and director of Women in Transition Seminars. She is also the co-director of the international project "Children as Teachers of Peace," and co-founder of the AIDS Hotline for Kids.

Norman Cousins was adjunct professor in the School of Medicine at the University of California, Los Angeles. He had previously been the editor, for thirty-five years of *Saturday Review* magazine. The author of 25 books, Cousins received numerous honors for his humanitarian efforts, especially those on behalf of world peace. He died in November 1990.

William Foege, M.D. was the executive director of the Carter Center, Inc., the Carter Center of Emory University, Global 2000, and the Task Force for Child Survival. He is a former president of the American Public Health Association and was director of the Centers for Disease Control from 1977 to 1983. He is currently Presidential Distinguished Professor of International Health at the Rollins School of Public Health at Emory University.

Jane Goodall spent 30 years studying and recording the lives of the chimpanzees at Gombe. Through her writing, lectures and National Geographic television specials, she has made it possible for an entire generation to grow up with the chimpanzees of Gombe and, in a larger sense, to find a new solidarity with wildlife and a heightened concern for the world's resources.

Erich Graesser studied Protestant theology with Rudolf Bultmann and Werner Georg Kummel. He became a full professor at Ruhr University in 1965, and has been a professor of the New Testament at the University of Bonn since 1979. Among his books is *Albert Schweitzer as Theologian*. His interests and research focus on New Testament studies, the ethics and theology of Schweitzer, creation ethics and the ethics of animal rights.

Benjamin Hooks has served as the executive director of the National Association for the Advancement of Colored people since 1977. An ordained minister, as well as a lawyer, he was the first Black member of the Federal Communications Commission.

Jackson Lee Ice, professor of religion at Florida State University and a former President of the Society for the Philosophy of Religion, died in August 1991, one year after the International Schweitzer Colloquium. His writings on Schweitzer included the book *Albert Schweitzer, Prophet of Radical Theology* and articles such as "Did Schweitzer Believe in God?," "What Schweitzer Believed About Jesus," and "The Deeper Meaning of Reverence for Life."

Homer A. Jack was a co-founder, with Norman Cousins and others, of SANE, and was its executive director from 1960 to 1964. A Unitarian Universalist clergyman, writer and activist for peace and civil rights, he is the author of *Albert Schweitzer on Nuclear War and Peace*. He died in 1995.

Gerald G. Jampolsky, M.D., psychiatrist, lecturer and author, has gained international recognition for his work with children who have catastrophic illness. In 1975 he established the Center for Attitudinal Healing in Tiburon, CA. He is the founder of the Children as Teachers of Peace project and co-founder of the AIDS Hotline for Kids.

John A.M. Karefa-Smart, M.D., physician and professor of medicine, is a former assistant director-general of the World Health Organization, and a former member of parliament of Sierra Leone, he is a member of the board of directors of the Albert Schweitzer Fellowship.

Antje-Bultman Lemke, professor emeritus in the School of Information Studies at Syracuse University, has lectured extensively on ethics, human rights and Albert Schweitzer. A consultant to universities and libraries on four continents, she translated a new edition of Schweitzer's *Out of My Life and Thought,* published in 1990. She is a member of The Albert Schweitzer Fellowship.

Stephen Lewis was a media commentator on public issues and a prominent labor relations arbitrator before becoming ambassador of Canada to the United Nations in 1984. In May 1986, at the U.N. Special Sessions of Africa, Lewis chaired the committee that drafted the five-year U.N. Program on African Economic Recovery. He is currently a special advisor to the U.N. on African issues, and a special representative for UNICEF.

George N. Marshall, a Unitarian Universalist Minister Emeritus, corresponded with Albert Schweitzer for many years before visiting him. On three occasions, at Lambarene. He is the author of the books *Albert Schweitzer: A Biography* (with David Poling) and *An Understanding of Albert Schweitzer,* and is a recipient of the Freedom House Award of Merit. He died in February 1993.

David C. Miller, M.D., was for many years associated with the Centers for Disease Control. In 1960 he spent six months at the Schweitzer Hospital in Lambarene, returning in 1965 to consult in the care of Albert Schweitzer during his final illness. Miller worked with his wife, Rhena Schweitzer Miller, on healthcare programs in a number of third world countries. He was the vice-president of the Albert Schweitzer Institute and co-editor of this volume. He died in March 1997.

Russell A. Mittermeier is president of Conservation International. Prior to this he served as vice-president of science at the World Wildlife Fund. In addition to numerous professional affiliations, Dr. Mittermeier is the author of five books, and dozens of scientific papers and popular articles on primates, reptiles, tropical forests and biodiversity.

Robert Muller retired in 1986 as assistant secretary-general of the United Nations. He is currently a consultant to the secretary-general and the chancellor of the University for Peace in Costa Rica. A recipient of the UNESCO Prize for Peace Education, Muller has written numerous books and lectures frequently.

Linus Pauling, research professor at the Linus Pauling Institute of Science and Medicine, was awarded the Nobel Prize twice: for Chemistry in 1954, and for Peace in 1962. He is the author of hundreds of scientific papers and several books, one of which, *The Nature of the Chemical Bond,* was one of the most cited scientific books of the 20th century. He died in 1994.

Ija Pawlowska is a professor of ethics at the University of Lodz (Poland) and the author of more than 150 papers in the fields of ethics and metaethics. She authored a book on Albert Schweitzer, organized a Schweitzer exhibit at the University of Lodz in the 100th year of his birth, and has lectured on Schweitzer both at home and abroad.

Victor W. Sidel, M.D. is distinguished university professor of social medicine of the Montefiore Medical Center, Albert Einstein College of Medicine, New York City. He was a founding member of Physicians for Social Responsibility, and served as its president in 1987. In 1984-85, he was president of the American Public Health Association.

Dennis Weaver, the popular actor best known for his portrayals of Chester in *Gunsmoke* and Sam McCloud in *McCloud,* is also an active environmentalist, currently on the board of directors of the Earth Communications Office. In addition, he is the founder of LIFE (Love is Feeding Everyone) and a national spokesman for CARE.

About the Editors

David C. Miller, M.D., formerly with the Centers for Disease Control, first met Albert Schweitzer in 1960, while conducting cardiovascular research at the Schweitzer Hospital and in the surrounding area. Dr. Miller and his wife, Rhena Schweitzer Miller, worked together on nutritional surveys and projects of primary health care in a number of third world countries. They served together on the executive board of the Albert Schweitzer Institute in Wallingford, CT where Rhena Schweitzer Miller now serves as honorary president. Dr. Miller died in March 1997.

James Pouilliard is a writer, editor and publisher. He served as a member of the Executive Board of the Albert Schweitzer Institute in Wallingford, CT and the editor and principal writer of the Institute's quarterly newsletter, *Legacy.*

About the Institute

The Albert Schweitzer Institute is a nonprofit organization that conducts educational programs in the U.S. and abroad in health care development and youth ethics. The Institute conducts interactive conferences that explore clinical, public health, human rights and ethical issues in order to stimulate health care initiatives that improve the lives of underserved populations abroad. The Institute also initiates educational projects that teach young people ethical values and encourage their commitment to community service in the United States and abroad. Programs are inspired by Albert Schweitzer's exemplary humanitarian service and philosophy of "Reverence for Life."

Founded in 1984 as the Albert Schweitzer Memorial Foundation, the Institute's offices are located on the campus of Choate Rosemary Hall in Wallingford, Connecticut. Over the years, the Institute has organized conferences, lectures and workshops, sponsored fellowships and awards, shipped humanitarian aid to developing countries, and publicly advocated for human rights and world peace.

Health Care Development Program:

The Institute's main program activities in health care development are the Schweitzer Conferences and Seminars (SCS), a series of dynamic meetings for health care leaders held in Central and Eastern Europe and the former Soviet Union (CEE/FSU). These interactive conferences aim to improve the region's ability to respond to the unique health needs of its vulnerable populations. Conferences explore public health issues of critical importance to the region, including cancer, children's health, HIV/AIDS, nursing, palliative care, physical and mental disabilities, prison health, refugee health, reproductive health, and tuberculosis. Over 60 of these "Schweitzer Seminars" have been organized since 1994 by the Institute in collaboration with the Open Society Institute's network of offices in CEE/FSU, making SCS one of the most extensive health education series operating in the region.

Humanitarian Aid and Youth Service-Learning Program:
The Institute, in collaboration with Recovered Medical Equipment For the Developing World (REMEDY) and Healing the Children Northeast, has developed an innovative youth service-learning program called *Ethics in Action Through Humanitarian Aid*. Initiated in 1999, this program uses the experience of working on a humanitarian aid project, combined with participation in workshops to discuss the physical, social, political and economic conditions of the receiving country and the issues surrounding shipment and receipt. Critical thinking skills are enhanced and students learn why and how to participate in, or even initiate humanitarian aid programs in their own communities.

Youth Ethics Education Program:
As a preventive response to the growing problem of youth violence and intolerance, the Institute and The School for Ethical Education (SEE) have developed a series of youth workshops called *Building Ethical Communities Through Service-Learning*. Begun in 1998, the interactive, day-long events bring together diverse groups of Connecticut elementary, middle and high school students, teachers and parents to discover shared community values, learn and practice ethical leadership and cooperative work skills, and design and implement student-led community improvement projects. Following the workshops, the Institute and SEE staff provide ongoing guidance to student-led groups to assist them in completing successful projects. Workshop structure and curriculum materials are being designed to make it practical for individuals and organizations throughout the country to implement this educational model in their own communities.

With the help of experienced secondary school teachers, additional curriculum supplements called *Schweitzer Action Packs* are being edited and tested. Based on the writings and experiences of Albert Schweitzer, students Kindergarten through Grade 12 are challenged to reflect upon their own situations in this world and to conceive of actions and activities they can initiate to support global health.

Publications / Website:
The Institute encourages efforts to reissue and distribute works by and about Albert Schweitzer and has partnered with Johns Hopkins University Press to reprint four of Dr. Schweitzer's most important books: *Out of My Life and Thought, The Primeval Forest, The Quest for the Historical Jesus,* and *The Mysticism of Paul the Apostle.*

On the web, www.schweitzerinstitute.org provides information about the Albert Schweitzer Institute and its current program activities. Alternatively, information may be obtained by contacting the Institute at P.O. Box 550, Wallingford, CT 06492. Tel: 203-697-3933, Fax: 203-697-3943.